Exploring the Mystery of Matter

Exploring the Mystery of Matter

The ATLAS Experiment

Production and Photography by Claudia Marcelloni
Written by Kerry-Jane Lowery
Scientific Editing by Kenway Smith

Production team

Production Manager, Creative Director and Photographer: Claudia Marcelloni
Written by Kerry-Jane Lowery
Copy Editor: Neal Hartman
Scientific Editorial Committee: Kenway Smith (senior editor); Manuela Cirilli and Heinz Pernegger (assistant editors)
Conceptual Designer: Fabienne Marcastel
Production Designer: André-Pierre Olivier

Thanks

The production team would like to offer their sincere thanks to the ATLAS Management for commissioning this work, and to all of the members of the ATLAS Collaboration for their support in preparing it, designed as a tribute to their collective efforts. In particular we wish to express our gratitude to the many people, too numerous to mention individually, who contributed their recollections and anecdotes in interviews, offered photographs and encouragement. Paul de Jong deserves a special mention for his contribution of the Glossary of terms, which we hope will assist non-physicist readers. Martine Desnyder-Ivesdal is also thanked for her contribution to the proof-reading.

First published in Great Britain in 2008 by Papadakis Publisher

PAPADAKIS

An imprint of New Architecture Group Ltd.

Head Office: Kimber, Winterbourne
Berkshire, RG20 8AN, UK
www.papadakis.net

ISBN: 978 1 901092 95 0

A CIP catalogue of this book is available from the British Library.

Printed in China.

The future begins now

In the 16th century, Copernicus forever changed man's perception of his place in the cosmos. Since those times, our understanding of 'what is out there' has been challenged as our knowledge has grown. At the end of the 20th century we found ourselves in a similar position to our forebears towards the end of the 19th. We thought the Theory of Everything was around the corner when in fact a new warehouse full of unexplained phenomena awaited us. With ATLAS coming of age, we may gain the rights of passage we so eagerly await at the beginning of this new millennium.

› *Superconducting magnets in the 27 km Large Hadron Collider tunnel, which straddles the French-Swiss border on the outskirts of Geneva.*

Foreword

ATLAS, and the other big LHC (Large Hadron Collider) detectors, are technological and sociological marvels, which I hope and expect to produce marvellous scientific results.

The idea of putting a large hadron accelerator or collider in the LEP tunnel at CERN was conceived in the late 1970s, some years before LEP itself was approved. The technological basis of the collider itself was clear from the beginning, although very challenging development was needed to reach the desired performance, which is far beyond anything achieved in the past.

However, whether detectors could be built that could handle the very high particle production rates at the LHC was at first completely unclear. A vigorous R&D programme at CERN, launched in the early 1990s, seems to have resolved the problems by pushing the technologies to the limit. Indeed there is now such confidence that ATLAS and the other detectors will work as expected that people are already discussing an upgrade of the LHC that could increase the data rate tenfold!

The first discussions, in the 1970s, of the idea of forming a global collaboration to build and exploit accelerators assumed that the partners would be Europe, the USA, Russia and Japan. The much larger number of countries that, thanks to CERN's open door policy, contribute to the LHC experiments is astonishing. Involvement has boosted the level of science and technology in many of these countries, and the diversity of ATLAS makes it an exciting place to work.

The scale and sophistication of ATLAS are amazing, and it is inspiring to witness scientists from around the world working together at ATLAS to probe the nature of matter. This beautiful book conveys the sense of awe that strikes all those who visit ATLAS and meet the ATLAS team.

Professor Sir Chris Llewellyn Smith FRS
Director General of CERN 1994-98

Peter, "Official leak" 11/16/95

The LHCC recomends the approval of the ATLAS + CMS projects, together with the plans, including milestones, leading to the subsystem Technical Design Reports

The second prize is to get to build it.

Bonne Chance

Good continuation until the final success!

› A hand-written letter from the chief referee of the CERN Peer Review Committee for the LHC experiments (LHCC), gives ATLAS members the long awaited news that many years of preparatory work designing and defending the experiment have succeeded.

Preface

With the start-up of the LHC and ATLAS a great dream becomes an exciting reality! It was in 1984 when some of us started dreaming about a decisive experiment to hunt for the Higgs Boson and to discover SUSY. In our enthusiasm and ambition to explore a new territory of Physics, how naïve we were about what this really meant; nobody imagined the complexity and technical challenges ahead, culminating finally in the LHC and ATLAS. Even when ATLAS was born, with its Letter of Intent in October 1992, we were still a long way away from any realistic plan for the project.

This book captures some of the many great moments and special events and activities on the long way of a unique collaborative and human effort that led to this marvelous, technically sophisticated detector as well as to a not less amazing organization of numerous structured teams and working groups preparing eagerly for the data and physics analysis.

Building up the world-wide ATLAS Collaboration, based on the trust from all the partners, and securing the massive resources from many Universities, Laboratories and Funding Agencies all over the globe, was an extraordinary human adventure as well. In spite of all thinkable differences in culture and history, we fortunately all share the same motivation to explore the New Physics promised by the LHC. But without the generous support from all Funding Agencies, and from CERN as our Host Laboratory, we would never have been able to achieve our common goal.

The collective work on the ATLAS experiment started more than 15 years ago, and our thoughts today are also with all our colleagues who sadly could not see its final realization. Building ATLAS has been a great adventure, and we must all be grateful to have had the privilege to devote a large part of our lives to it. And yet, the best part is still to come, so let's remain curious as to what Nature will reveal and teach us!

Peter Jenni

Peter Jenni
Spokesperson, ATLAS Collaboration

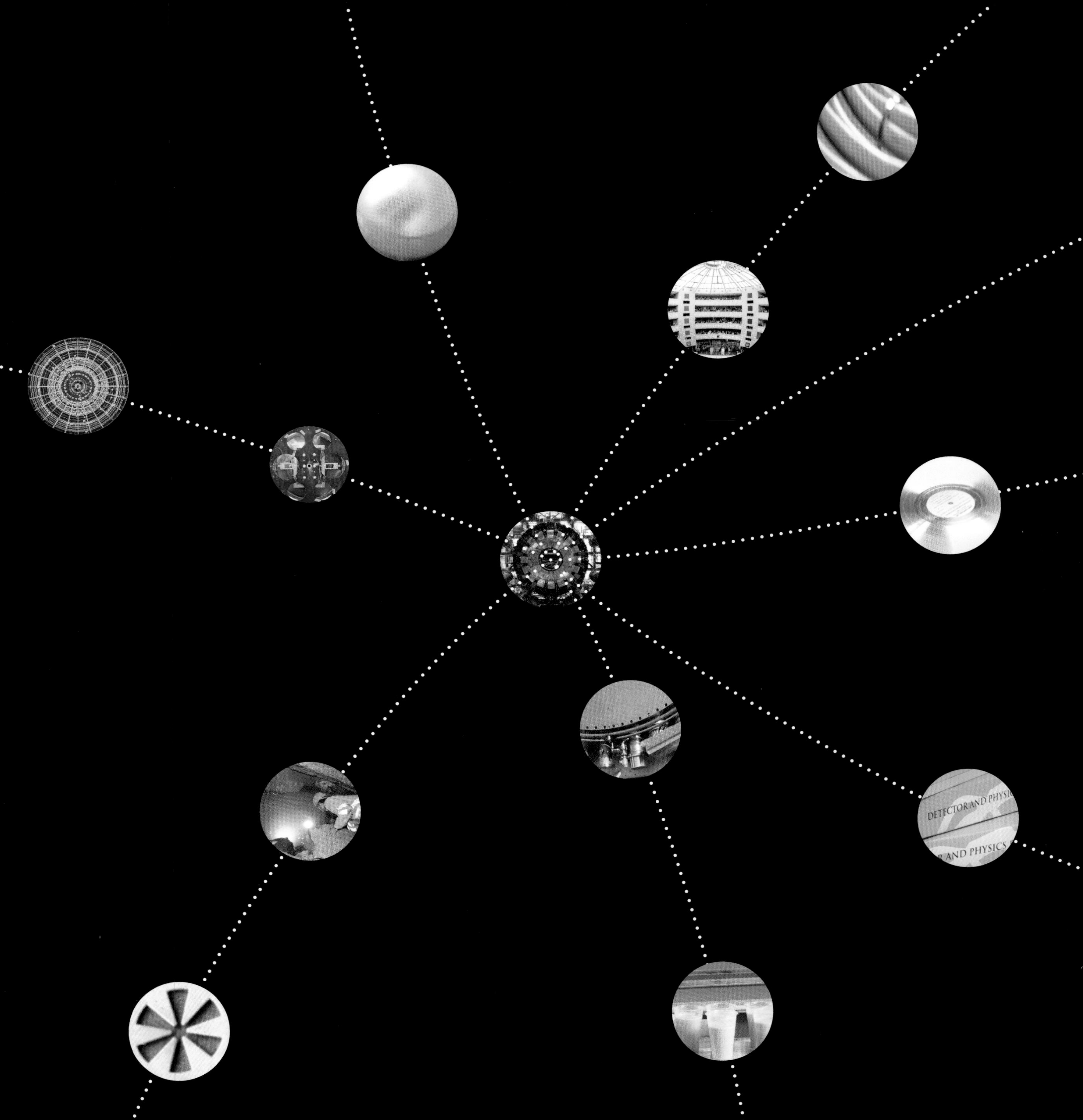
DETECTOR AND PHYSIC
AND PHYSICS

Contents

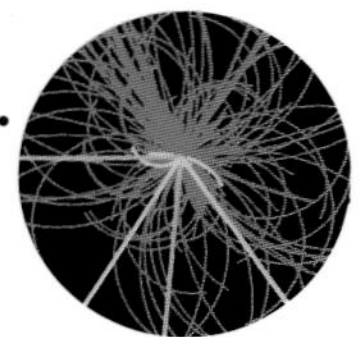

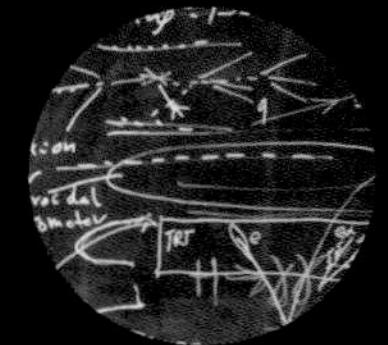

"This was a time of new beginnings"

Tables strewn with papers, coffee stains on scribbled formulae, no cigarette butts in ashtrays since this is a No Smoking area, a feverish and strained atmosphere – almost intoxicating. A small group of men and women sit in large armchairs debating the origins of time late into the night; some are lost in thought, others dashing off notes. All are fired up. Most great ideas begin in small circles and often these circles form around the dining room table. The birth of ATLAS was no different. The seeds were sown in CERN's Restaurant 1, back when there were plush armchairs and shorter queues.

Rewinding the Universe

Cast your mind back to 1989, to the fall of the Berlin wall, the end of the Cold War, the discovery of Neptune's three moons, the protests in Tiananmen Square, the Dalai Lama's Nobel Peace Prize, and to a lesser known event, the convergence of the 'particles' that would eventually form ATLAS. This was a time of new beginnings. To be truthful, it was not that romantic a beginning. Meetings in Barcelona and Evian had come first, the strategy discussed and decided upon and the collaboration carefully honed before the real physics started. The long talks in the CERN cafeteria were quick to follow. Now, let's go back in time.

The shape of things to come

From the very start of ATLAS the physics goals were clear, but the path leading to them was not. It was only in the very early nineties, when the first programmes of Research and Development were underway, that solutions to the detector's most fundamental challenges could be guessed at. Its shape would remain a mystery until certain elements, in particular the magnets, were decided upon. The experiment has four superconducting magnets, the largest number ever used in a single experiment. From the moment the choice of magnets had been made, ATLAS took on its characteristic shape. It also took on its name. And it took to collecting superlatives.

› *The detector's birth certificate, the Letter of Intent, was submitted in October 1992 by the ATLAS Collaboration to the LHCC, responsible for LHC experiment proposals.*

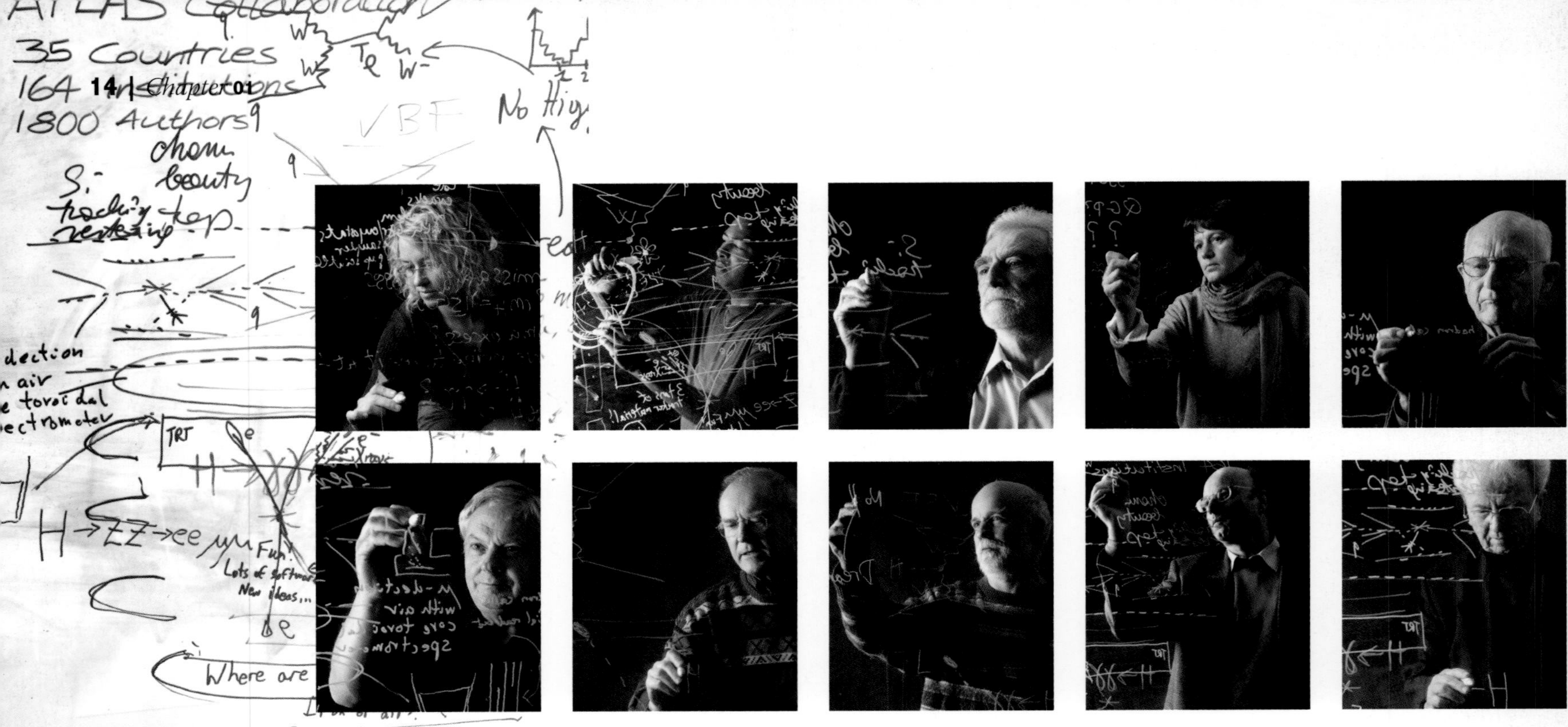

› *Physicists are big fans of blackboards. Here a host of collaborators sketch out the detector's origins.*

Standard model of perfection

One of the main purposes of ATLAS is to find the elusive Higgs. Without it the foundations of particle physics remain uncertain.

The Standard Model, the triumph of theoretical particle physics in the 1970s, describes three of the four known fundamental interactions between the elementary particles that make up all matter. It has successfully predicted the outcome of a large variety of experimental tests and is the grounding for particle physics. However, the model still lacks one crucial ingredient: we do not know what causes the fundamental particles to have different masses. Enter Higgs.

As far back as 1964 Peter Higgs, and independently François Englert and Robert Brout, proposed a mechanism whereby certain particles have no mass and others do, by postulating that mass is due to interactions with a new 'field', coined the 'Higgs field'. The related particle, the Higgs boson, is the missing link in the Standard Model.

There is no common understanding as to what the Higgs may be. Some theorists believe there is one Higgs boson, which would lead to a fairly simple theory - but this is physics so simplicity remains relative. The majority predict, and hope for, a Higgs mechanism requiring a more complex theory; and others think there may be something completely different awaiting us which would deal a lethal blow to the Standard Model. Such experimental demolitions of major theoretical predictions have historically sown the seeds for significant advances in our understanding of the Universe, by stimulating creative new thinking. Many hope for something wholly unexpected.

ATLAS Collaboration
35 Countries
164 Institutions
1800 Authors
charm
beauty
top
VBF
No theor
No Hig
Dream
H
TRT
Xrays
3 tons of tracker material!
H → ZZ → ee μμ
Lots of software
ith air
ore toroidal
spectrometer

A giant leap for mankind

Enter the LHC and ATLAS. At long last an accelerator powerful enough to bring forth the Higgs and a detector subtle enough to observe it. What will emerge remains a mystery. Beyond the Higgs, the physics community will focus on finding dark matter and supersymmetry, as well as extra dimensions. ATLAS's sister experiment, the Compact Muon Solenoid (CMS), will explore similar issues, whereas the LHCb and ALICE will address more specific questions of heavy flavour and quark-gluon plasma physics, respectively. The search is on for a stable and heavy particle, left over from the Big Bang, which is a component of dark matter. ATLAS is sensitive enough to indirectly detect weakly interacting particles; with a bit of luck they will be supersymmetric as predicted by some theorists. Such discoveries would reveal what a further 25 percent of the Universe is made of.

Coming of age

In the world of particle physics, projects take on lives of their own, extending over decades. ATLAS will be entering adulthood by the time it is 'switched on.' Eighteen years in the making, eighteen years to enter finally into new territory and take physics where it has never been before. Its life will be spent transcending frontiers, taking us to the origins of time and to the edge of the known Universe. It will explore the fundamental nature of matter and energy by creating conditions similar to those just after the Big Bang. This feat of engineering and logistics – it is one of the most complex experiments ever undertaken by mankind – will reveal the infant cosmos to us.

It is hard to believe that we only know what four percent of the Universe is made up of, while the rest remains a mystery. So the probability of discovering something fundamentally new, which could call into question everything physics has postulated so far, is very high. Although one declared purpose of the experiment is to find the Higgs, ATLAS may also reveal how galaxies are formed, what dark matter is and whether the Universe is expanding or contracting – even its ultimate fate.

02

"Never before has this been done"

Going Global

ATLAS is a project of epic proportions, spanning the globe, with Europe, North and South America and Asia forming the largest contingent, while Africa and Australia are less prominently represented. It is one of the first times ever that such a colossal scientific project has been executed on such a multi-lateral basis.

The level of faith individuals, institutes and governments have placed in ATLAS from the onset is quite unique. Its collaborators are passionate about their work and their dream. Over 2000 individuals from 170 institutions make up the Collaboration, and it is still growing as the detector approaches completion and the anticipated physics results become increasingly tantalising.

The sheer size of just about everything in ATLAS is mind-boggling – from the smallest ever to the largest ever. Initially, the project was divided into manageable sub-projects, each with a dedicated team of collaborators and a tale to accompany it. The many stories around these geographically and culturally diverse teams forming, growing, supporting and reinforcing one another and ultimately fusing into a single coherent ATLAS Collaboration are inspiring. May the tale begin.

› *Each institutional member of ATLAS has the right to vote on different matters related to the Collaboration in the Collaboration Board's sessions.*

Pulling Power

Institutions from far and wide have gravitated towards ATLAS, wanting to play a part in this epic endeavour. At all stages of the experiment institutes have joined the Collaboration, bringing with them precious knowledge, manpower, passion and funds to make the dream come true.

Exploring the fundamental nature of matter and the basic forces that shape our Universe requires commitment, innovation and funding. ATLAS has brought together a wealth of people, institutes and universities. A host of collaborators, from professors to students, engineers and technicians have worked alongside each other for the past 18 years to build this time machine.

The price to pay

One hundred and seventy institutions from 37 countries form this truly global Collaboration. The threshold is high to enter ATLAS, since the right mix of experience and resources (human and financial) is vital for the experiment to work. Among the conditions to join is the 'membership fee' fixed at 137'500 Swiss Francs. In addition to this fee, participants contribute to the Common Fund, set up to pay for elements which are too demanding for any one institute to build, such as the toroidal magnets, which cost one-third of the experiment's total budget. Over almost two decades of development, fluctuations in national economies and currency exchange rates have also created funding headaches.

Home grown

Alongside these cash contributions, each institute has either built part of the detector in its home laboratory, outsourced prototypes to be built by its national industries, and/or contributed manpower and expertise to the parts being assembled at CERN. At the national level, countries seek a balance between the scientific prestige gained by participating directly in an epic such as ATLAS and the economic benefits of winning subcontracts, some payable from the national funding contributions, others won through international competitive tendering. In other words, funds pledged to ATLAS can be spent at home and then the components sent on to CERN. At the university level, the research and development programmes benefit students and professors hugely, and for companies, the technology contracts are of great monetary value, and a chance to showcase the countries' industries.

› *ATLAS is synonymous with global scientific cooperation on an unprecedented scale. Here Japanese, Chinese and Israelis collaborate on the Muon Spectrometer Thin Gap Chambers (TGCs).*

Building Blocks

Collaboration is the cornerstone of ATLAS. Without it the detector would still be the figment of quite a few imaginations. But it is now a reality, as well as one of the greatest and most successful international collaborations in the world of science.

Where else but in ATLAS can you find Japanese, Russians, Chinese, Israelis and Pakistanis working together on the same project, with Americans, Europeans and others from 37 countries? For the past 15 years or so, the Japanese and Israelis have been working together on the muon spectrometer TGCs ('Thin Gap Chambers' for mere mortals). Their teamwork goes back further in time, but as far as ATLAS is concerned the mid-nineties was their moment. Starting off with two institutes, one in each country, now over ten are involved.

Simply does it

This collaboration began quite simply, really. A few colleagues from one institute spoke with a few from another over the waters; the discussions were shared with the upper echelons; travel ensued between Israel and Japan; meetings were set up at CERN to shape the collaboration as a whole, while smaller ones took place in the home institutes to work out the details; then followed the development of chambers for ATLAS. Easy!

It took time to develop the right type of chambers, of course, divide the work, and keep everyone happy. Finally it was agreed that the various groups involved should work on every aspect of the building of the chambers, so that the knowledge was shared by all. The first laboratory for construction was built in Israel, where the two countries worked together, while in the meantime the Japanese prepared theirs and were ready to roll a year down the road.

Three's company

And this is where it gets interesting. Around this time, a new player arrived on the block. The Chinese also wanted a piece of ATLAS. The turn of the century was to be auspicious. Two institutes in China were keen to build TGCs, and after a number of discussions with the Israelis it was agreed to give ten percent of the work to China. Counterparts from China came over to learn how to build the chambers, whilst at home a new laboratory was built to take the construction back there. The Israelis gave all the material and the tooling to the Chinese in exchange for their manpower - a fair trade if ever there was one. The chambers were then tested in Israel and Japan and sent on to ATLAS. The collaboration is still ongoing since technicians from the three countries came over to install the chambers in their final 'place of rest' in ATLAS.

Overall it was a very challenging time, involving cultural differences, language barriers, distances between the countries and laboratories, and the training of people. It was also a learning curve like no other for most involved, and quite simply an extraordinary and unexpected collaboration.

was a thrilling as well as a tough period”

100% Design

The design phase of most projects tends to be one of the most exciting and creative periods. This is when inventiveness comes into its own, when various possibilities are considered and developed and the best one 'wins'.

ATLAS began on a blackboard and graduated to prototypes, which became bigger and bigger as time went by and decisions were made. Brainstorming was rife during these times, the wildest ideas allowed and considered with only a few seeing the light of day.

It was a thrilling as well as a tough period since making choices based on performance, cost and schedule means that an individual or a group's brainchild will be chosen over that of another. Such times are never easy for those who must abandon the technology they have been researching and developing for a number of years, but this is the name of the game. In essence, ATLAS is the outcome of a mixed marriage. Compromise and the best interest of the experiment are at the root of this endeavour.

› *The ATLAS Technical Design Report (TDR). Every aspect of one of the largest experiments in the world is carefully documented and peer reviewed.*

A

In the ATLAS Internal Note GEN-NO-002, dated 18th March 1993, a specially convened ATLAS Magnet Panel recommended to the Collaboration one of three options being considered for the Muon Spectrometer magnet configuration. This was a decisive moment, since the final shape of ATLAS would be dictated by the choice made.

Magnetism

The selection of a suitable magnetic field configuration to measure the paths of charged particles largely defines the dimensions and cost of a detector and is crucial to its performance. In the past, typical collider experiments have only had one magnet system; ATLAS has four: a solenoid magnet for the Inner Tracker, a Barrel Toroid Magnet and two Endcap Toroids. The paths of charged particles produced in collisions at the LHC bend within these magnetic fields, enabling the measurement of their charge and momentum. But why choose a relatively costly air-cored toroidal magnet?

Independence day

"The magnet panel unanimously recommends option 1: all superconducting air-core toroids," reads the Internal Note. The recommendation was made in spite of the larger expense involved, on the basis of the superior physics performance. One of the main advantages of opting for air-cored toroids is that the Muon Spectrometer is able to operate 'standalone'; its performance does not depend on the tracking information coming from the inner detector. It is like having two detectors in one, each providing an independent and complementary measurement of a muon's trajectory.

Bigger, better, faster, more

ATLAS collects superlatives, so this chapter has one too. The Superconducting Barrel Toroid is the largest magnet ever built, which means that the magnetic forces acting on the coils are among the most powerful ever dealt with. When the coils are fully driven, the force on one metre of superconducting cable is comparable to the power of a jumbo jet engine. This has required that the coils be embedded in a suitably strong framework to avoid the detector jetting off.

The challenges are numerous when building such a complex and groundbreaking detector. Using a toroidal field in a collider experiment was the first step into the unknown; many more were to follow, including the sheer size and magnitude of ATLAS. Just think, the energy stored in the magnets could lift the 7000 tons of the detector itself, or the Eiffel Tower, for that matter.

› The first of the eight 100-tonne Barrel Toroid Magnet coils is painstakingly lowered and manoeuvred into its final position with centimetre accuracy.

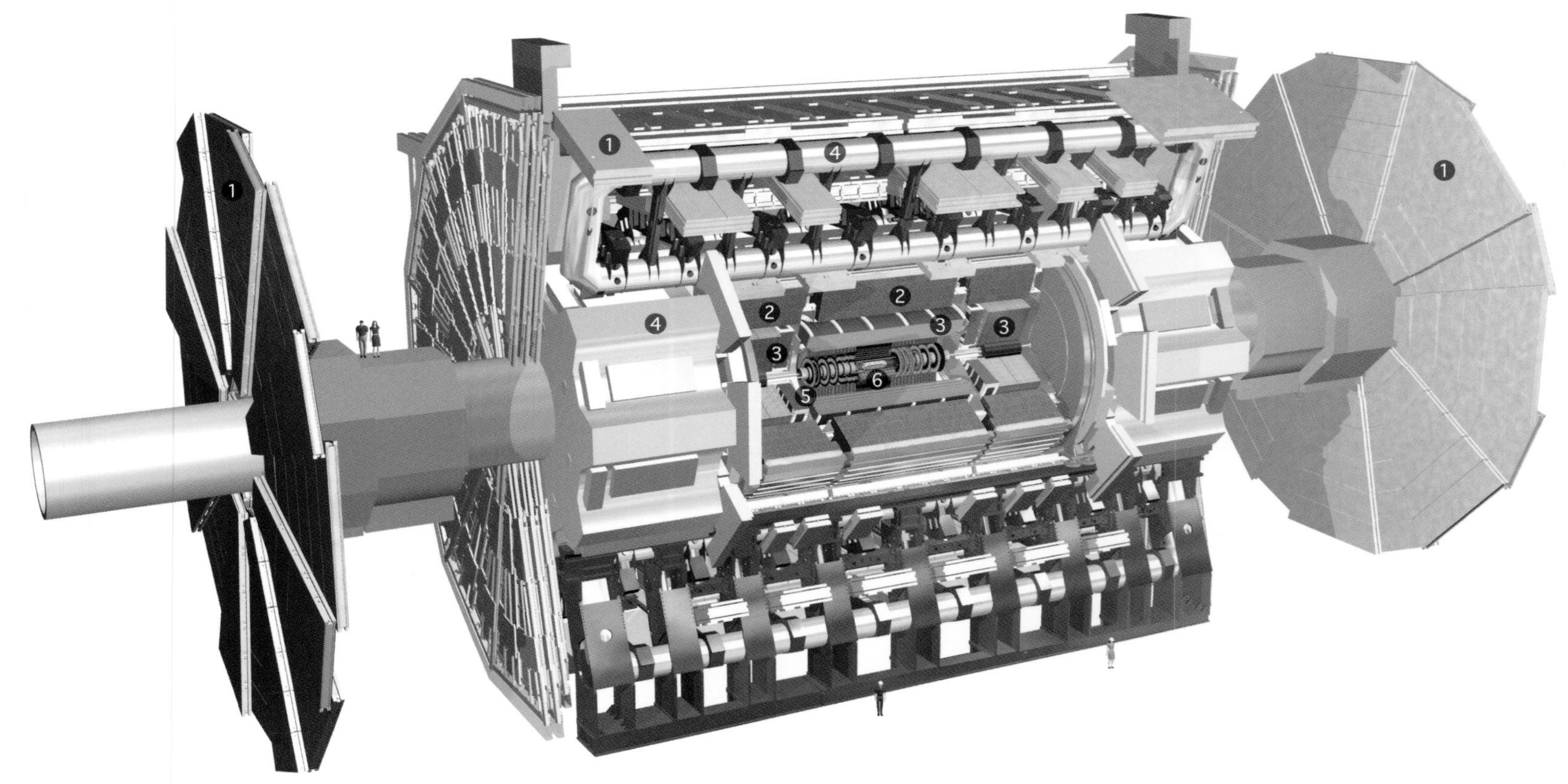

› *Cutaway of the ATLAS detector with its main subsystems:*
1 - Muon Detectors
2 - Tile Calorimeter
3 - Liquid Argon Calorimeter
4 - Toroid Magnets
5 - Solenoid Magnet
6 - Inner Detector

Catwalking

ATLAS quite unique in that it has an air-core toroid, which allows for a lot of space – everything being relative that is – inside the detector. So much so in fact that for the first time it is possible to enter certain areas of a functioning detector and work inside it. There are over half a kilometre of catwalks throughout ATLAS enabling physicists, engineers and technicians to fine-tune, repair and clean (and strut their stuff) whenever necessary. This has of course led to a new challenge as far as security issues are concerned, since it is difficult to know if there is anyone inside when the beam is turned on. Tight and futuristic security developed by ATLAS will help.

Gaseous Nobility

Noble gases and high-energy physics go hand in hand. ATLAS has its own favourite.

Liquid Argon is tasteless, colourless, odorless, non-corrosive, non-toxic, non-flammable, extremely cold and largely inert to boot. Overall, not the most exciting of gases. So it is no surprise that its name, derived from Greek, means 'the lazy one'.

Argon makes up approximately one percent of the Earth's atmosphere and belongs to the family of rare inert gases. It was the first noble gas to be discovered and is the most plentiful of them, which is partly why it was chosen for the ATLAS calorimeter. Noble gases rarely react with other elements since they are already stable, and are therefore ideal candidates for high-energy physics.

Two other noble gases with far more interesting names, Xenon and Krypton, were considered for the calorimeter, but neither was suitable. The former is very rare and expensive (though it is used sparingly in the transition radiation tracker), and the latter, although cheaper and more available, can lead to purification problems (and is fatal to Superman).

A liquid Krypton calorimeter would have improved its energy resolution and increased the sensitivity to a possible Higgs signal by ten percent, but it would also have increased the overall running cost by seven million Swiss Francs. So, in the end, ATLAS favoured Argon, the noble commoner, which, luckily for us all, is harmless to our favourite superhero.

The photographs on the following pages are of models and prototypes from the Research and Development period.

› ***Inner Detector*** - A prototype wheel of the Transition Radiation Tracker (TRT), overall comprising 296 km of four millimetre diameter carbon fibre straws. Below an SCT prototype module.

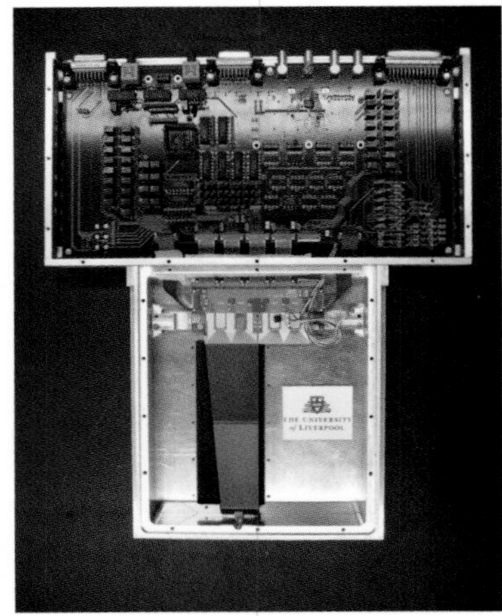

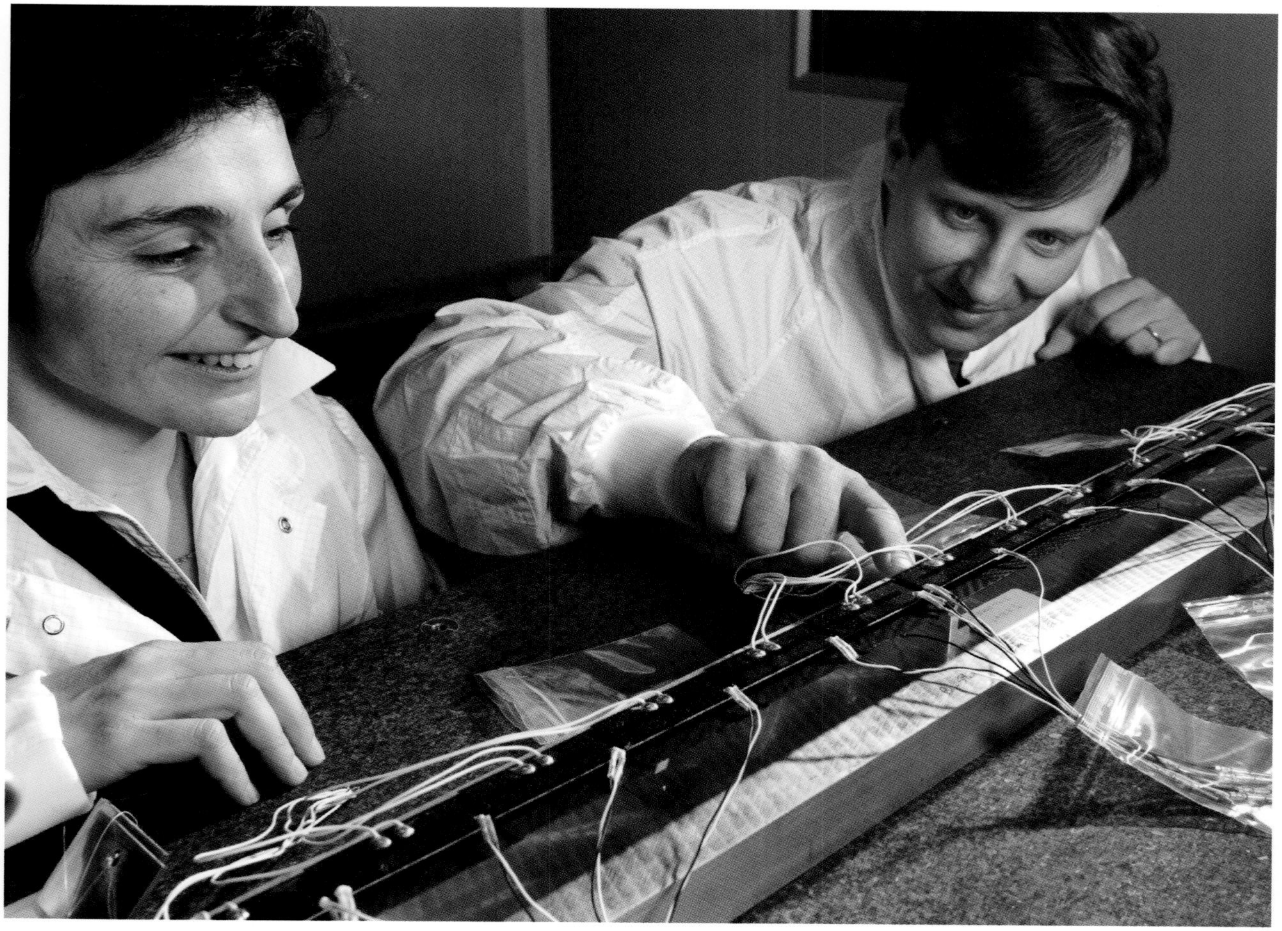

› ***Inner Detector*** - Pixels account for more than 90 percent of the Inner Detector's 86 million channels, but occupy just 1.5 percent of its volume.

› *Liquid Argon Calorimeter*

The 'accordion' calorimeter is composed of 4000 electrode-absorber layers in ultra pure liquid Argon to maximise stability, granularity and speed.

› ***Tile Calorimeter*** - Half a million pieces of blue scintillator plastic, 187’000 pieces of green fibre and half a million plates of steel make up the multi-layered tile calorimeter.

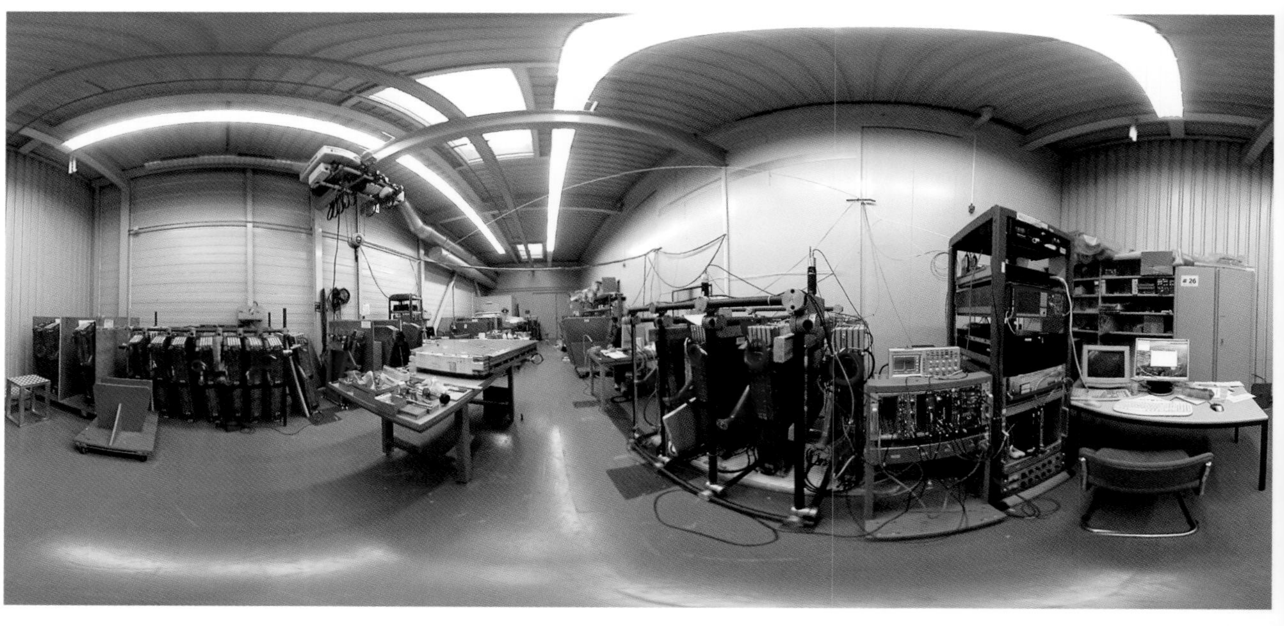

› ***Muons*** - The 4000 individual muon chambers are built using four different technologies by 48 institutions in 23 production sites around the world.

› ***Magnets*** - The Barrel and Endcap Toroid superconducting magnet together form the largest electromagnetic volume ever created by mankind. Below an endcap toroid model.

04

"What's Monte Carlo got to do with it?"

Monte Carlo or Bust

A glass falls to the ground and shatters into a million pieces. Two test cars crash head-on at high speed, debris everywhere. Protons collide in a detector, a shower of particles results. This is indeed physics, not the latest James Bond movie. No two shattered glasses, car accidents or particle collisions are the same. There are myriad trajectory variations each shard, car component or particle can take, each with its own probability.

Simulation is the tool that allows physicists to predict collisions where a large number of probabilities are at play, and Monte Carlo is at the heart of it. No simulation can be said to perfectly reflect reality; it is only when physicists start taking data that they can compare what really happens with their predictions and theories. Let's head to Monte Carlo to get to the bottom of this.

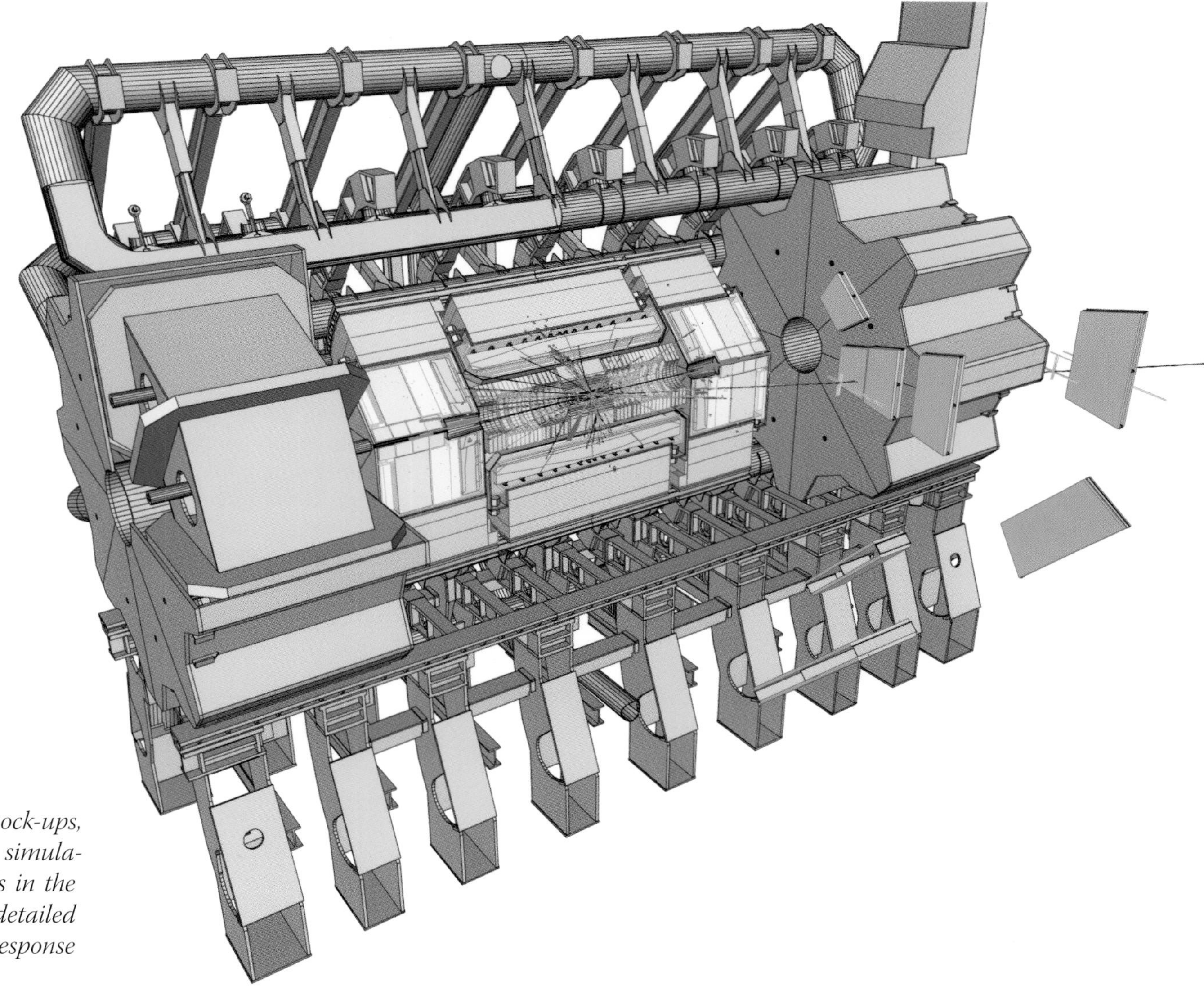

› *Models, whether physical mock-ups, computer graphics or detailed simulation software, are essential tools in the design of ATLAS, providing a detailed understanding of the detector's response to different physics processes.*

Dicing with Debris

What better place to simulate than in Monte Carlo? Sounds more glamorous than it actually is. When physicists get excited about working with Monte Carlo and mention 'simulation' in the same breath they are in fact talking about a computational algorithm invented at Los Alamos National Laboratory in the 1940s.

The Monte Carlo method is a mathematical technique widely used in banking, economics and computer games. It simulates the behaviour of a complex process using probability and a large quantity of randomly generated numbers to create an approximate solution. It is very much like throwing multi-faceted dice to figure out all the possible combinations or figuring out the most likely trajectory of a free kick in football depending on, for example, the angle of the foot, the strength of the kick and the friction of the grass.

Lightning fast

When protons are accelerated in CERN's Large Hadron Collider the particle beams are steered to collide in the middle of the ATLAS detector. Examination of the elementary particle debris from these collisions reveals information about fundamental physical processes. Superconducting magnets guide the beams, consisting of approximately 3000 bunches of 100 billion particles each, as they circle the 27 kilometre ring. No two collisions are the same: different particles and trajectories will be produced each time and unstable emerging particles will decay into different particle combinations. With so many possibilities, physicists need a way to understand the data they take.

Virtual reality

Simulation is used from the onset when designing a detector in order to optimise its conception and later on to mimic what happens in a working detector. It is an extremely important part of the experiment since its aim is to generate data that is as close as possible to the real data that will come out of the detector once it is running, allowing for something interesting in the real data to be identified more easily. The development of the reconstruction programme relied completely on simulation, only to be proven when we go from the virtual detector to the real one.

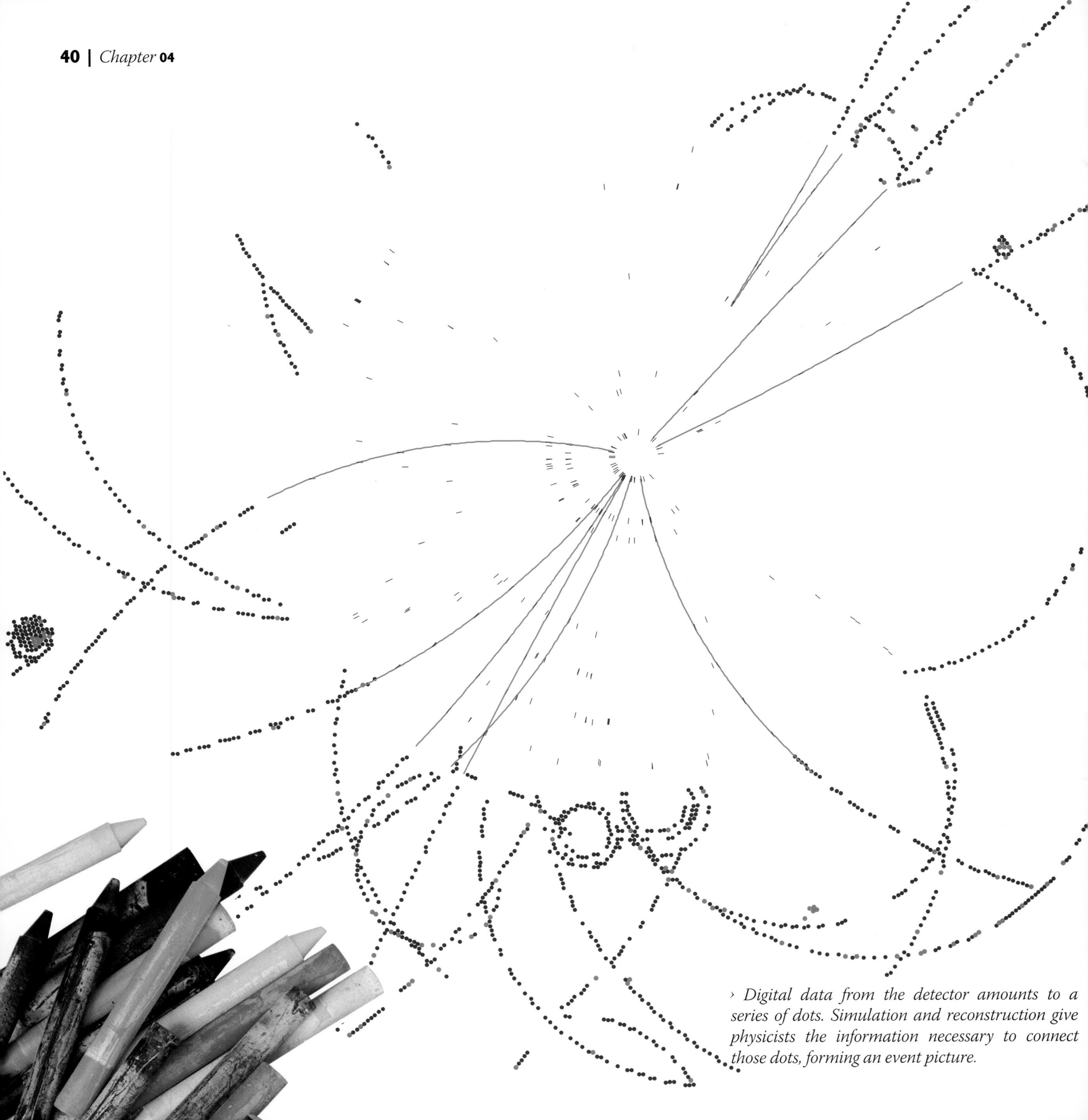

› Digital data from the detector amounts to a series of dots. Simulation and reconstruction give physicists the information necessary to connect those dots, forming an event picture.

A foreseen-unknown

So with simulation telling us what should be obtained from the detector according to the theorists' predictions, physicists know where to look in the hundred million or so digitized electronic signals for indications of events of specific significance. The particle trajectories must then be reconstructed by putting the electronic messages back together, giving the physicists data to analyse and hopefully the chance to find something new.

Each reconstruction is unique, belonging to a particular event and time, and can lead to multiple analyses depending on what specific physics is being looked for. The whole Collaboration will be intent on extracting as much information as possible from these events, keeping an eye out for the unusual. This is what physics is all about, the predicted unknown.

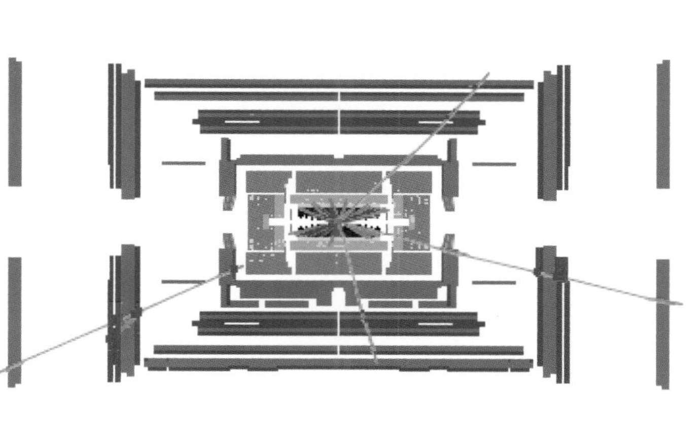

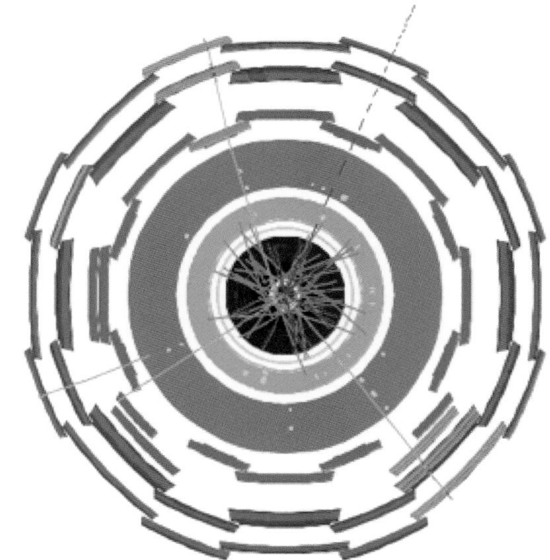

› A simulated Higgs boson event, showing transverse and longitudinal cross-sections of ATLAS, with 4 charged tracks from the Higgs decay emerging from the collision point and reaching the outer layers of the muon chambers.

05

"It's like panning the ocean for oysters to find a few pearls"

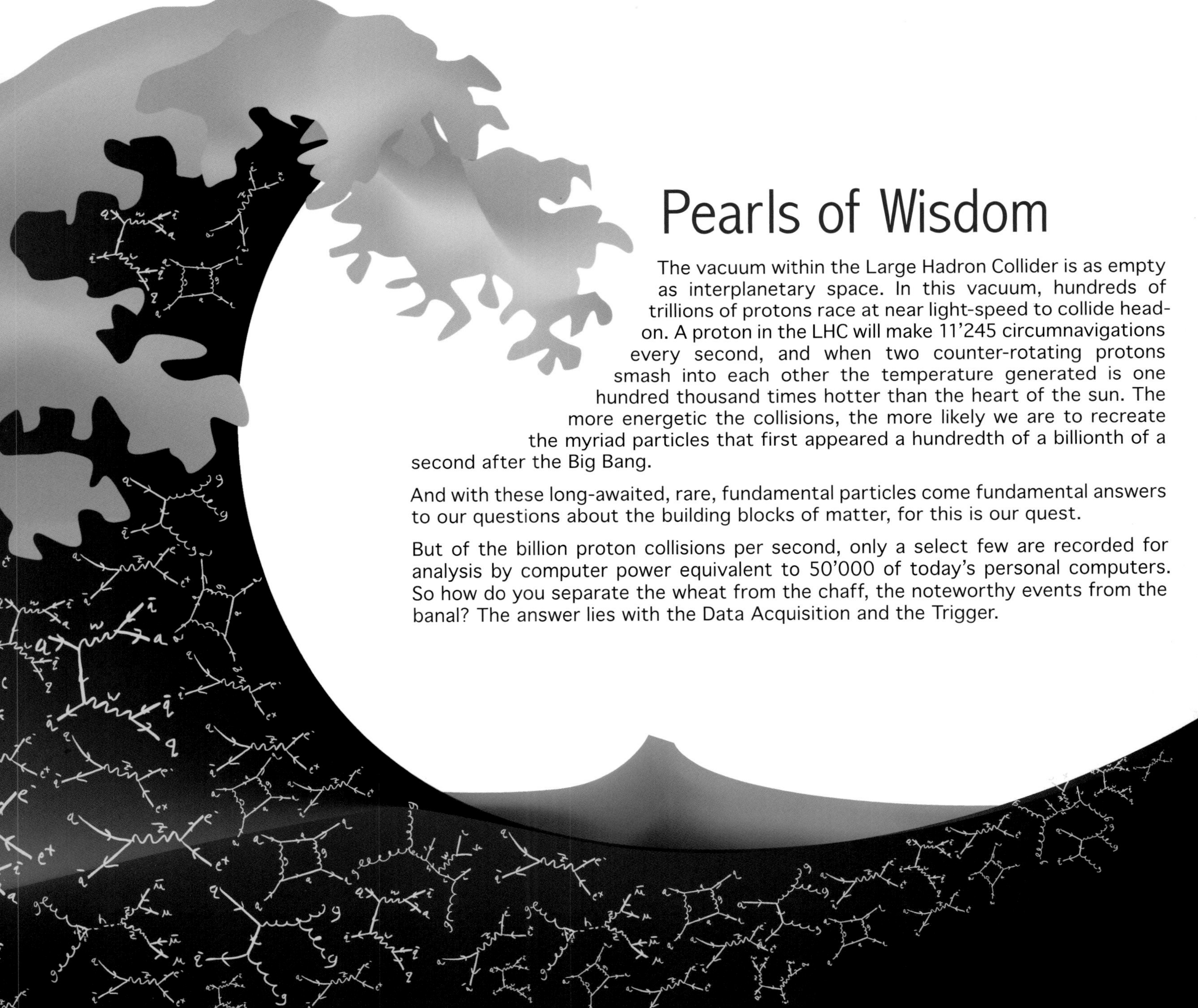

Pearls of Wisdom

The vacuum within the Large Hadron Collider is as empty as interplanetary space. In this vacuum, hundreds of trillions of protons race at near light-speed to collide head-on. A proton in the LHC will make 11'245 circumnavigations every second, and when two counter-rotating protons smash into each other the temperature generated is one hundred thousand times hotter than the heart of the sun. The more energetic the collisions, the more likely we are to recreate the myriad particles that first appeared a hundredth of a billionth of a second after the Big Bang.

And with these long-awaited, rare, fundamental particles come fundamental answers to our questions about the building blocks of matter, for this is our quest.

But of the billion proton collisions per second, only a select few are recorded for analysis by computer power equivalent to 50'000 of today's personal computers. So how do you separate the wheat from the chaff, the noteworthy events from the banal? The answer lies with the Data Acquisition and the Trigger.

Straining for the Less Ordinary

Sampling and recording the debris from 40 million events every second with approximately 25 protons from either side crashing into each other is a Herculean task. Especially when you add into the equation the fact that 40 million events would require one million CDs to record the data flowing from them all - every second, that is. Only then do you come to realise that the selection process is rather vital and time is of the essence.

A definition to start with:

An event is a digitized electronic 'snapshot' of what happens when two bunches of protons cross and some collide. There are nearly one billion of these collisions every second. The flow of data from them is managed by the Data Acquisition (DAQ) and Trigger, the hardware and software which together select and record in real time and for subsequent analysis the rare and new processes - from the bulk of uninteresting interactions - that physicists are searching for.

To simplify things, let us imagine that the DAQ is pushing event data through the Trigger, a rather large and sophisticated sieve.

Sifting along

Forty million events pour into one side of this sieve every second and 200 - two hundred only – come out the other end and are recorded for posterity. From this ocean of information all we want are a few pearls of wisdom. This is quite a sieve. In order to achieve this goal the sieve has three levels: Level-1 Trigger, Level-2 Trigger and the Event Filter. Level-1 is fast: it has 2.5 millionths of a second to decide for each event whether to send it to the next level or flush it down the drain. Level-2 comprises around 500 computers which select 3000 events from the 100'000 received from Level-1 each second. A further 1900 computers complete the work, reducing the 3000 events to the mere 200 that flow from the Event Filter to the computer centre where the information is saved forever. By the end of the sifting process, the equivalent of only one compact disk is engraved every two seconds.

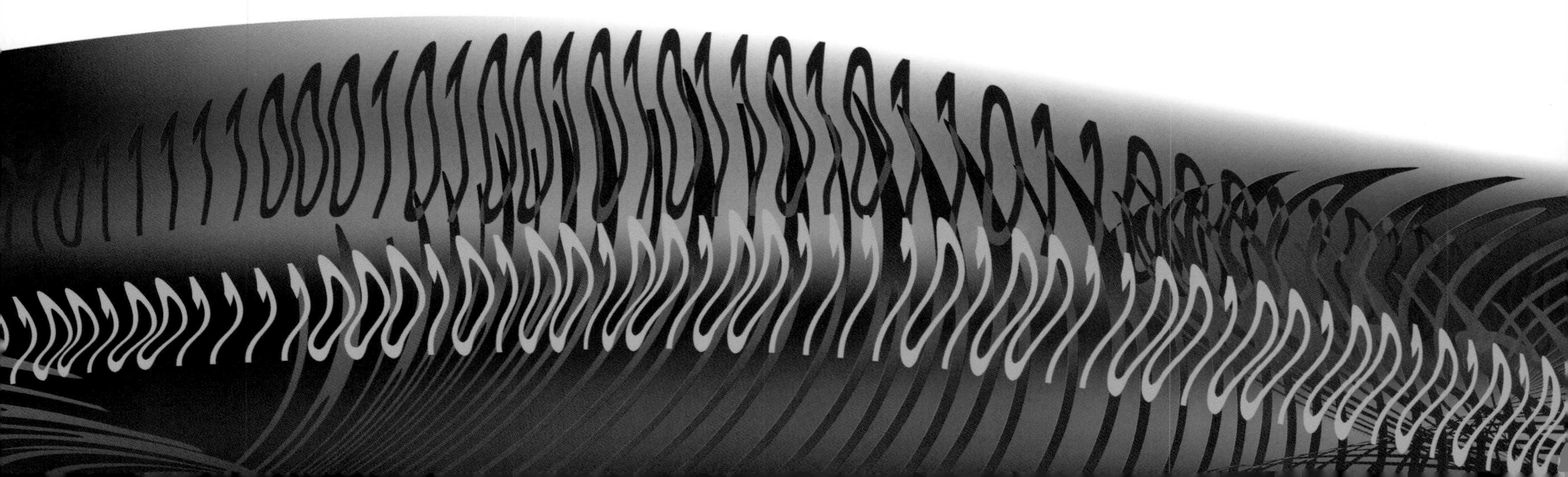

Throughout, the DAQ and Trigger communicate and work in parallel. While the Trigger is deciding what information to store, the DAQ collects all of the data from the detectors, waiting for the Trigger to reject or accept the information to be streamed to the next level.

Grading oysters

As complex as the DAQ and Trigger are, the events they strain from this ocean of information are still just raw data – a pile of oysters – most of which do not contain the rare pearls that are so eagerly sought after. It is up to the physicists to grade these oysters by which varieties are most likely to contain their silky gems – Akoya, South Sea, or Tahitian oysters – and then to determine the most efficient ways to find the most rare and beautiful examples. Physicists are not even all looking for the same types of oysters – it all depends on what particular processes they are interested in. Even after the oysters have been retrieved and sorted, it takes many examples and much more work – known as analysis – before the information leads to a new discovery.

The quest for the perfect pearl has only just begun.

› *At any given moment about 80 events can be found in the Level-1 Trigger, like a doctor having 80 patients constantly lingering in the waiting room.*

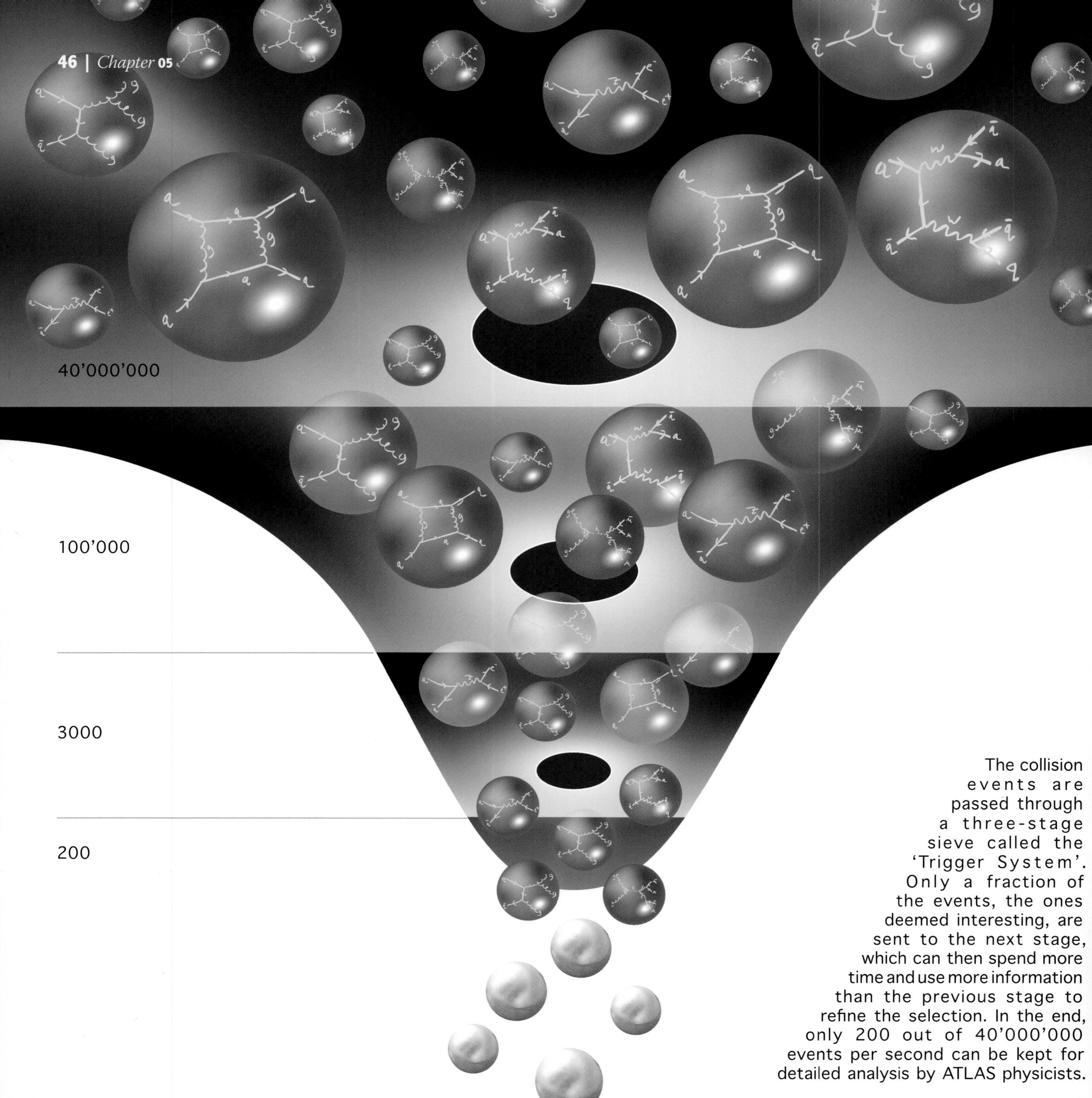

The collision events are passed through a three-stage sieve called the 'Trigger System'. Only a fraction of the events, the ones deemed interesting, are sent to the next stage, which can then spend more time and use more information than the previous stage to refine the selection. In the end, only 200 out of 40'000'000 events per second can be kept for detailed analysis by ATLAS physicists.

› ***DAQ - Trigger.*** The Event Builder puts together in a synchronised way the data from the independent sub-detectors, much like solving 2000 puzzles of 150 pieces each every second.

this way to the grid page 144

Please Hold

To comprehend the amount of raw data the detector produces let us go global. Six and a half billion human beings on the planet, every one of us placing seven phone calls simultaneously, that is how much raw data is produced by ATLAS. For the most part it is discarded without even being looked at. Following a collision the data to be processed proceeds along a plethora of paths. These thousands of optical fibres move the information to the Data Acquisition and Trigger (T/DAQ) for sorting, much like routing multiple phone calls to a single switchboard.

Thank you for waiting

The first stage of the T/DAQ must deal with four billion bits of information every second. Currently, a call on a mobile phone sends 10'000 bits of information per second, so the information coming from ATLAS is equivalent to 400 million phone calls being placed simultaneously, every second, each and every one channelled through the same operator.

Because there is so much information the T/DAQ uses thousands of computing units to analyse parts of the tracks left by the particles, or fragments of conversations left by the callers. It is only at the end of the selection process that the entirety of an event – or the whole of a conversation – is stored in one computer for future analysis. Then the Grid kicks in.

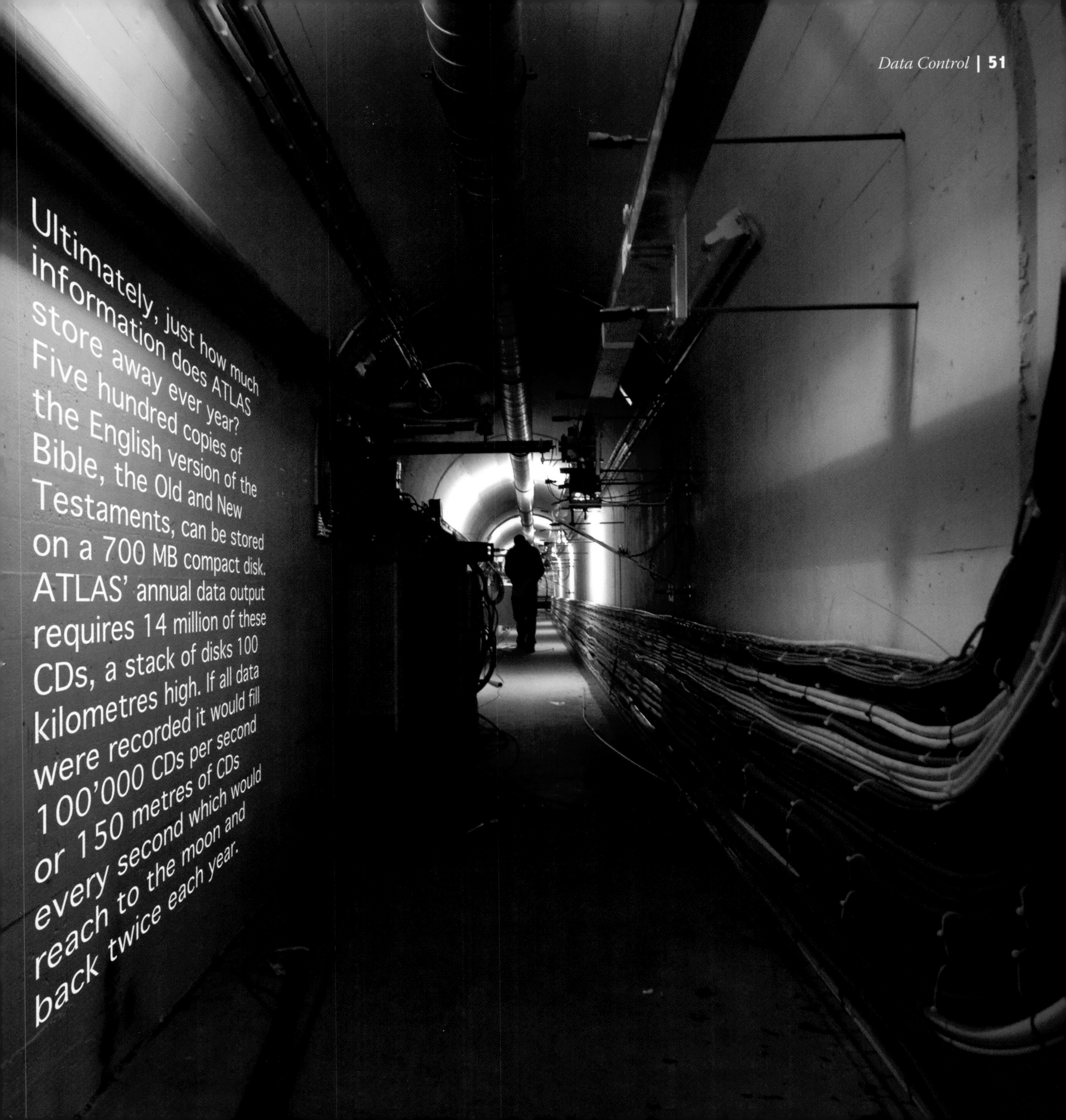

Ultimately, just how much information does ATLAS store away ever year? Five hundred copies of the English version of the Bible, the Old and New Testaments, can be stored on a 700 MB compact disk. ATLAS' annual data output requires 14 million of these CDs, a stack of disks 100 kilometres high. If all data were recorded it would fill 100'000 CDs per second or 150 metres of CDs every second which would reach to the moon and back twice each year.

06

"It's quite a miracle of engineering, inventiveness and sweat"

Inventing the Wheel

ATLAS was not built in a day, nor was it constructed entirely by the physics community. Various industries were called upon to make this colossal endeavour a reality. At times, prototypes developed within science laboratories were contracted to companies to produce the pieces in large quantities. At other times, the science community worked alongside specific industries to develop the necessary components or tools for the advancement of the experiment.

The physics community has surpassed itself in building ATLAS and some of the companies who have risen to the challenge have been awarded the ATLAS Supplier Award for their work and dedication. Inventions have also abounded during the construction phase - extremely clever and innovative creations - the 'brainchildren' of some of the most gifted and able physicists, engineers and technicians of our times.

› *'Daisy petals' embedded in kapton film provide an elegant solution for electric connections. The kapton acts as a printed circuit carrying the high voltage between petals.*

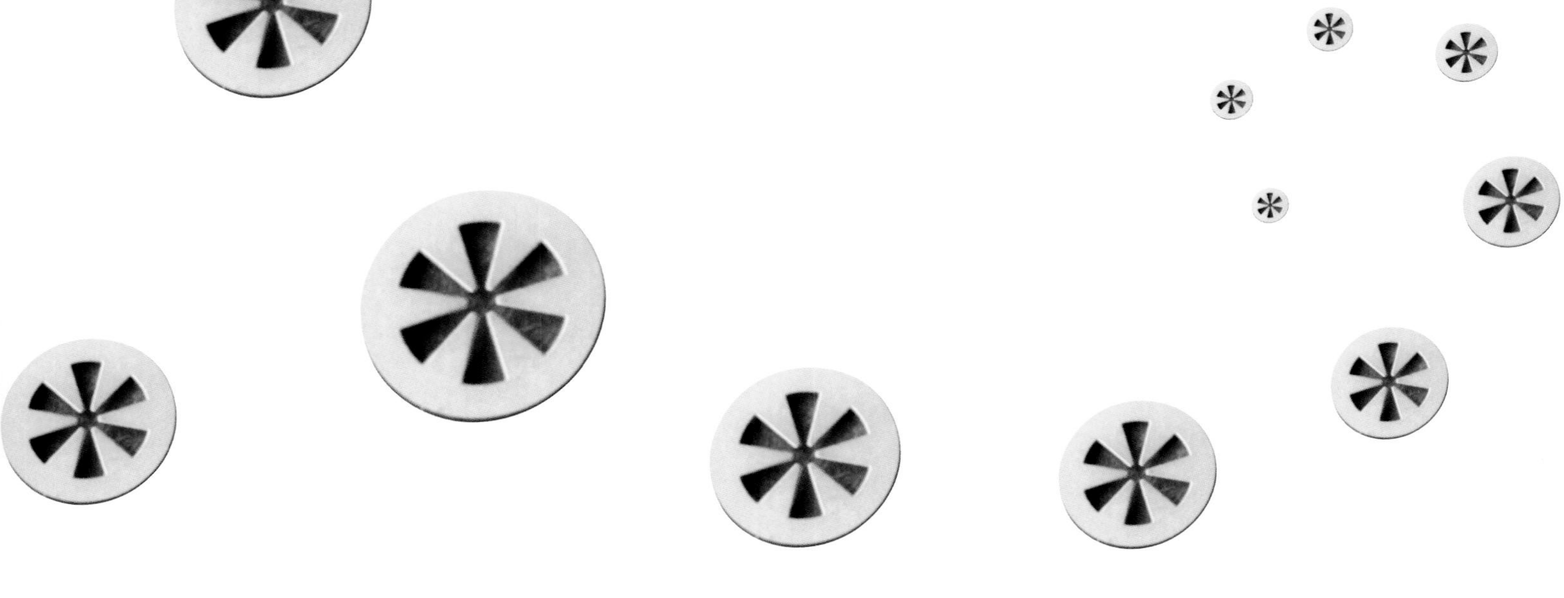

The Idea Factory

ATLAS is a whirlwind of inventions. Here's one right now: high response speed + stability + good granularity + no cracks = accordion shaped calorimeter. There you go, one of the most cunning ideas in the world of modern detectors.

Inventing is one of the many things people involved in physics excel at. They do it daily because that is what is expected of them and as the saying goes: 'An invention a day keeps the particles at play.' So, before the detector is even switched on it will already have stretched and challenged some of the greatest minds in science and given 'birth' to a host of new ideas and possibilities.

Spreading the gospel

Not satisfied with challenging its own people, the world of physics likes to make the outside world work that little bit harder too. Companies from all walks of life have taken the opportunity to work for the exacting world of particles. Indeed, working with ATLAS brings with it a certain prestige. There is also often a development side to the work, offering interesting prospects for companies eager to improve their technology with input from some of the world's leading minds. And when people are called upon to build something totally new and unusual in a field often viewed as mysterious and futuristic, motivation is high.

A rich harvest

There are numerous examples of companies outdoing themselves. In fact, the Collaboration has been so impressed by some of its suppliers' work, dedication and passion that it created the ATLAS Supplier Award. Companies from Holland, Japan, Byelorussia, Great Britain, France, Russia,

› *The accordion shape of the lead/stainless steel plates of the calorimeter was inspired by the musical instrument.*

› *The Supplier Award was created in 2001 to acknowledge companies that have gone above and beyond expectations.*

Israel, Switzerland, Norway, Pakistan, the Czech Republic, the United States, Slovenia, Germany and Italy are amongst those that have met the rigorous criteria necessary to win the award. The former owner of SB Verksted, a mechanical construction company in Norway, actually bought back his company when he realised that the new owners were not living up to the standards he had worked so hard for throughout his life. The damaged cryogenic tank ATLAS had been sent by the company was returned and repaired. It is now a part of the detector, as are the three other 50'000 litre tanks, which arrived on time and fully functioning. Needless to say SB Verksted won the award.

Pole position

The auto industry is replete with examples of high performing suppliers. One of carmaker Skoda's offshoots won an award for the forward shielding elements. And, just to make sure it will definitely be the fastest detector around, ATLAS commissioned a small firm that makes the front cones for McLaren's Formula 1 cars to build some of its carbon composite structures. Fast indeed. But wait, Mzor from Minsk, a leading specialist in tractors and tanks, built a number of elements for ATLAS too, including absorber plates for the Liquid Argon Hadronic Calorimeter. Tractors and tanks? Seems that there is room for all in the world of particle physics.

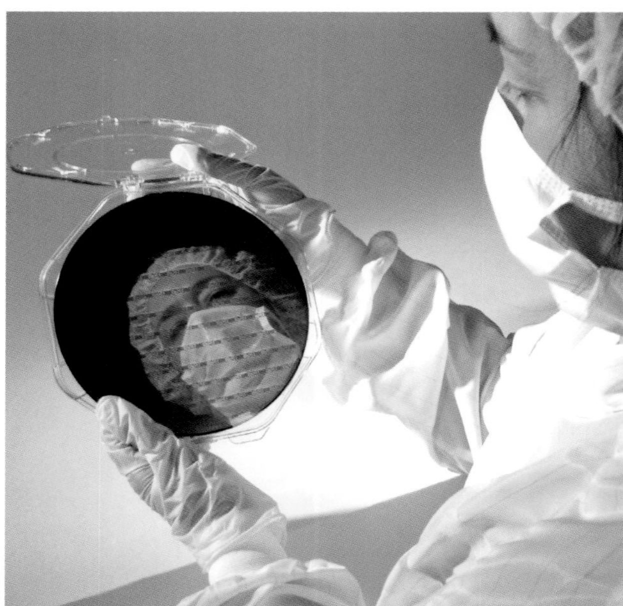

› *Inner Detector* - Production and assembly of the Inner Detector's structure and electronics takes place in clean facilities around the world.

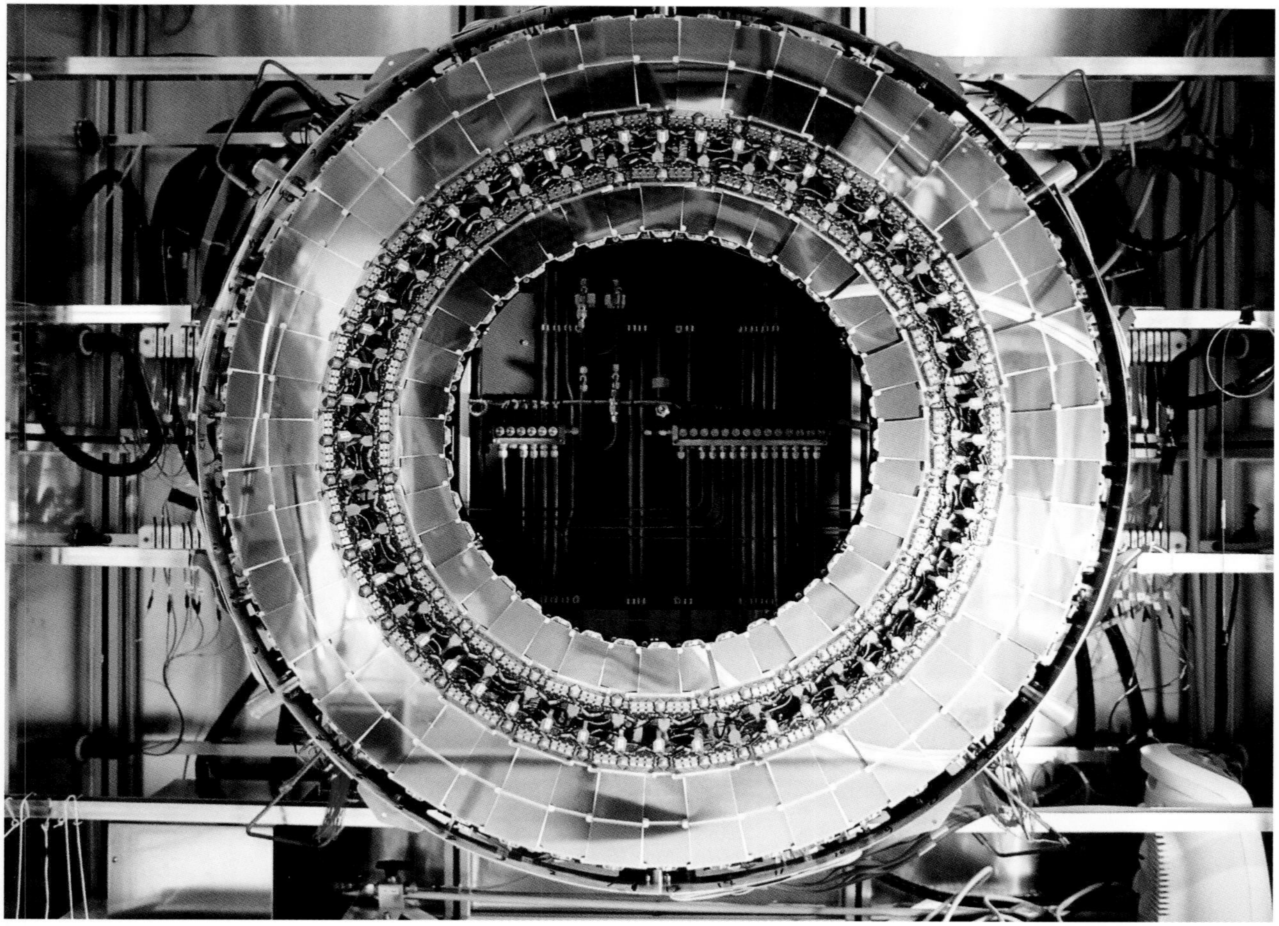

› *Liquid Argon Calorimeter*

For electron and photon identification it pays to stay really cool: 182 modules for a total of 600 tonnes of material are immersed in liquid Argon, at -185.9°C.

› ***Tile Calorimeter*** - Thousands of optical fibres that are individually checked to spot damage read out the flashes generated by particles showering in the detector.

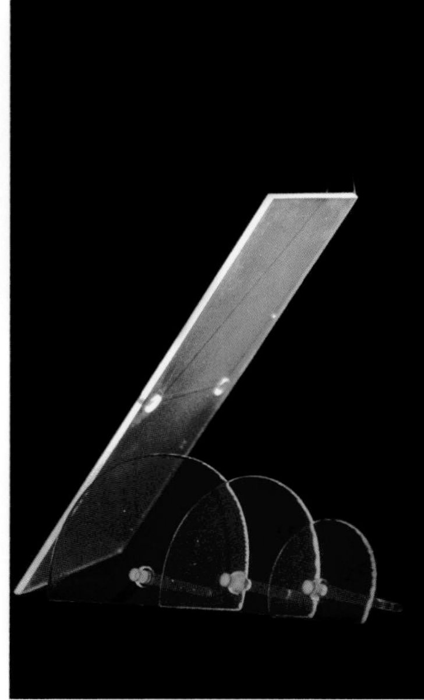

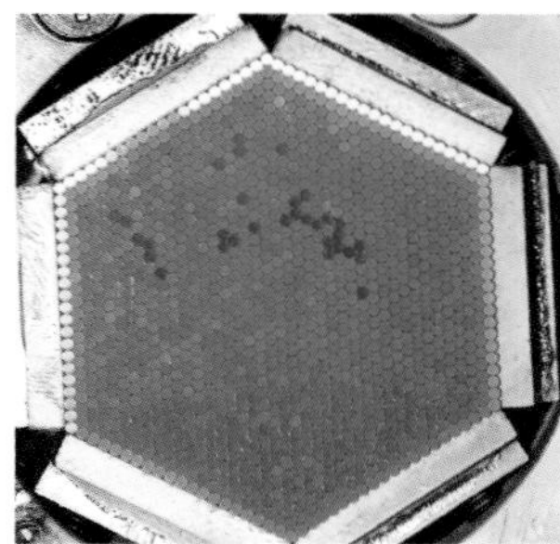

› ***Muons*** - MDT chambers are tomographically scanned to verify signal wire positions with 20 micron accuracy (approximately a fifth of the thickness of a human hair). All other muon technologies went through similar quality assurance procedures during the construction phase.

› ***Magnets*** - When fully powered, the energy in the largest of the magnet systems, the Barrel Toroid, is equivalent to 10'000 cars travelling at 70km per hour.

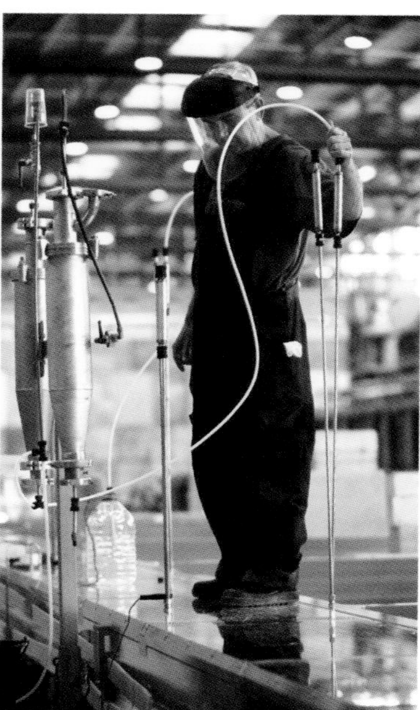

07

"There's a truth towards which we're converging"

The Ultimate in Melting Pots

CERN, the leading particle physics laboratory in the world. People from just about every continent collaborating, in situ or remotely, on the many projects underway. The buzz: the LHC, the most powerful scientific instrument on the planet. And finally, ATLAS and CMS, both the largest scientific experiments on the LHC and together, one of the biggest scientific collaborations in the world. This is the forefront – this is where the next era of particle physics begins.

On the human level a multiplicity of languages, cultures and nationalities. An international collaboration communicating in sign language at times. Post-docs participating in some of the most exciting physics in decades. Physicists, technicians and engineers travelling the globe. Professors teaching students in classrooms thousands of miles away via video link. Women negotiating their way in a world of men. When you peer through the looking glass, this is what the world of physics is about.

Show and Tell

"Shalom! Salam Alaikum! Buongiorno! Gutenmorgen! Morning! Bonjour!" We are in Switzerland, though it is hard to tell with the array of greetings going on around you. But then this is ATLAS. In one building alone there are teams from six different countries working together.

How can technicians from every corner of the planet work on the same project – where precision is of the essence – speak different languages, and yet get the work done? ATLAS is a cornucopia of nationalities where body language, facial expressions, drawings, international signs such as the 'thumbs up' and expressions like 'OK' rule.

Hakuna matata

"With people like this you can't have a problem. The communication just works, it's not an issue. We work together, sometimes eat together too, and take tea." So it is not only the variety of languages that fascinates; culture plays an important role too. The 'tea brewing' cultures share their love of leaves with the potato or wheat loving ones who come into their own later in the day. Teams from specific countries have taken to inviting others over for meals so as to share their national dishes and practise their sign language. ATLAS will have transcended borders and cultures before it is even 'switched on'.

Culture club

There is a real sense of camaraderie at ATLAS, a genuine interest in the other, and a tangible pride in working on such a large and meaningful project. "Every day we share something, and in the end we all make the same product and it's neither theirs, nor mine, but ours," an Israeli engineer explains enthusiastically.

The ATLAS environment is as apolitical as they come, with people working as individuals. "You meet people and cultures, not necessarily nationalities, which makes things very comfortable," an American physicist explains. The Collaboration thrives on the multiculturalism in its own ranks. And every morning for years to come you will be greeted by hellos in a variety of languages and responding no doubt with your newly learnt greeting in Urdu, Hebrew or Russian.

› *The Pakistani team brews tea at 11:00 sharp every morning in building 180. Their 'official tea maker' always makes enough to serve the many international takers.*

Only Six Women, but How Many Hats?

They receive peaches from collaborators from Armenia, sip black tea and eat salami in the offices of the Russian contingent, and go food shopping with a medley of physicists, engineers and technicians of different nationalities for the many parties they organise. Six women, lots of hats. Souvenir sellers, caterers, negotiators, real-estate agents, restaurant scouts, school advisers, tea ladies, handywomen - these are a few of the roles they take on. Oh, and they have their secretarial duties to squeeze in too!

Hello

The Secretariat is the first point of contact for most of the foreign collaborators who come to work at ATLAS. It can be their lifeline at times when their language fails to break down barriers at the doctor's or they just cannot seem to get that paperwork right for their visa. Six women with solutions at their fingertips.

You're welcome

Impossible is obsolete with them around. Without them it would be that much harder finding a school for your children; a few weddings would not have taken place; the baby wall in building 40 would not smile back at you with clear evidence of a different kind of ATLAS activity. Without them ATLAS would grind to a halt.

Apprenticeships in Frequent Flying

Fully fledged engineers, technicians and physicists travel a lot, as do their apprentices. Surprisingly, the World Wide Web has not greatly impacted on their travel, but then when the draw is a foreign conference on Deep Inelastic Scattering or Physics in Collision it is easy to understand the need to jump on a plane to travel the globe.

› *Students sitting in an auditorium in the United States interact with their professor giving a class in the small teaching centre at CERN.*

Despite the lack of music, the Vienna Wire Chamber Conference attracts a large following every year as do the main gatherings such as the European High Energy Physics Conference, the American Physical Society Meeting and the International Conference on Lepton and Photon Physics.

Transient work

These well-attended worldwide conferences attract their fair share of ATLAS collaborators. Add to this the need to see colleagues and work directly on the detector every few months, whether you are on the other side of the planet or just an hour's drive from Geneva, and you have a fair bit more travel right there. So when is there time to work? On the plane and in transit, obviously.

Remote control

To appease its jet set crowd, and reduce strain on the travel budget, ATLAS has implemented a state-of-the-art video conferencing system, and CERN has set up remote teaching facilities for physicists who must hold classes at their home universities while they work on site. The Remote Video Conference Room also serves as an outreach resource for communicating with other schools. All that is needed on the remote end is a portable teleconference centre and a computer, allowing the physicist to conduct the class and interact with the students in much the same way were they all physically present in one auditorium.

It's a Man's World

The world of physics is undisputedly male dominated. The ratio of men to women is approximately 10 to 1, but few are complaining loudly. In fact, women tend to find physics a good environment to work in. Granted, it is high pressure and time consuming, but they receive respect and equal opportunity.

Women approach things differently, bringing new or distinct outlooks. They talk more of physics in terms of beauty and perfection, using words like 'gorgeous' and 'elegant', imbuing the science with a sense of artistry.

Coupling

Many a relationship has fallen to the wayside in the world of physics. It engulfs its aficionados, relinquishing its hold on them only years later, richer in terms of universal knowledge, often poorer when it comes to family life.

Science does not wait for the pregnant or the sick; it strides on, taking its followers from one pursuit to the next, leaving the stragglers behind. Were there better childcare, more flexible kindergarten hours, longer assignments and perhaps a more philosophical approach to physics at school, maybe then women would be better represented in this world of men.

Exponential Knowledge Growth

Talk to Post-docs working on ATLAS and it becomes obvious that they have had to learn the ropes fast. Gone is the safety of university. They must make their own way at CERN and raise their game alone, making work contacts and new friends in the process.

"You have to communicate, share, ask, otherwise you're lost," a group of enthusiastic students explains. "You have no choice but to learn and learn more since it's all there for the taking."

Student varsity

It is the opportunity of a Post-doctoral physicist's lifetime to work at CERN since the leading experts are here and the learning curve is so steep. Hierarchy all but disappears when they are on site; they melt into the crowd and have access to those in the know. Furthermore, they become the focal point of their laboratories, and so are exposed to so much more than they would be at home.

Whether at CERN or on one of the many business trips to attend conferences, give papers and generally get their work out there, there is still the semblance of a student body, albeit a far more international and supportive one. Students from the various experiments help each other out professionally, go skiing on weekends and to jazz clubs during the week, recreating in their very own way a student life in a foreign land.

› *Students benefit from their supervisors' experience, and in turn the academics are kept abreast of developments at CERN by students based there.*

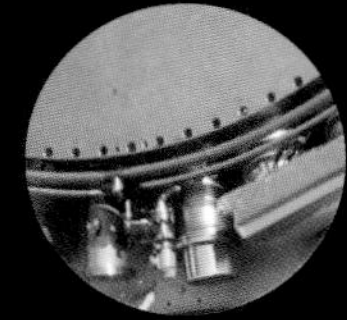

08

"Every institute brought its own piece to ATLAS"

Trains, Planes, Trucks, Prams and Automobiles

Just as ATLAS will take mankind on a voyage, so have its components journeyed from the four corners of the Earth to Meyrin, on the outskirts of Geneva, to become one colossal machine. The sheer size of many of the parts of ATLAS, as well as their extreme precision and fragility, have meant that travel has been at times challenging – and in some cases mind-bogglingly complex.

Bits of ATLAS have arrived in Geneva by train, plane, car, van, hydraulic truck, baby carriage and on foot. Since Switzerland is land locked, many pieces were first transported by sea or waterways and then transferred to special trucks to complete the pilgrimage. And what a journey it has been.

› *An Endcap Toroid Magnet is transported by a special 128 wheel hydraulic truck from the CERN main site to the cavern - 2 km apart - preventing it from tilting en route.*

Bridging the Gap

There are a myriad transport stories at ATLAS; everyone has one. Never mind the travel between continents, some of the shortest distances have proven to be the most testing. Just crossing the road from the main site of CERN to the ATLAS cavern has been problematic at times.

A passer-by was so fascinated by what he saw when the transfer of the Liquid Argon Cryostat between CERN and the cavern took place, that he drove his BMW straight into the bus in front of him. It has entered the annals as ATLAS's first collision. Then there was the Italian transporter whose delayed arrival was due to a pronunciation problem. His instructions were to go to 'le CERN', which for an Italian sounds like 'Lucerne', a town in central Switzerland. The rest is history.

Snail's pace

Two of the most epic journeys were those of the vessels for the Endcap Toroid Magnet. They were built in four shell-like pieces in a Dutch port, and transported, two thirty-tonne shells at a time, by barge and truck to Geneva. The itinerary was chosen with care since bridges and tunnels had to be avoided at all costs, along with train tracks – bar one crossing. The six metre wide, over six metre high and inordinately long convoy travelling at ten to fifteen kilometres an hour, with its six strong police escort, three company cars and four truckers, took five days to travel the 400 kilometres between Strasbourg and Geneva - including the dismantling of the high-tension lines between three and four in the morning when the Poligny train track had to be crossed.

Shelling bridges

However, the trip for the second two pieces proved to be the bigger challenge. The ski season was in full swing and a bridge had unexpectedly appeared in the Jura to enable skiers to cross over the road from their hotel to the slopes. After much deliberation - and France offering to build a side road that would cost ATLAS a mere 650'000 Euros - a cunning plan was devised involving a very large crane. One toroid shell at a time was lifted off the truck's trailer, which then sped under the bridge while the piece was swung over and redeposited on the other side. It took a full day to transfer both pieces and secure them to the trailers. The trip of a lifetime for all involved.

› *The convoy carrying the second Endcap Toroid vessel passes a mountain bridge in its own special way. Luckily the road is closed for the day.*

› *The first ATLAS collision – unfortunately between two local commuters distracted by the Barrel Cryostat crossing the road just outside CERN's main site.*

› *3000 tonnes of laminated iron was transported by train from the Tatra Mountains all over Europe. The wheels hadto be changed to fit the different tracks between countries.*

First Class Four-Wheel Drive

Fact: The heart of the detector arrived in a pram.

Definition: Pram - a four-wheeled carriage for a baby (Oxford Concise Dictionary).

Story: Once upon a time two young physicists wheeled a sporty all-terrain baby carriage containing neither baby nor teddy bear through the crowded security checkpoints at San Francisco International Airport, past the duty free shops, and down the never-ending corridors to the gate where their flight to Geneva was leaving from.

The carriage contained a box – a high-impact black plastic case to be precise – which was just slightly too big to fit into the airline Economy seat. Shame really. But it did fit snugly into the First Class armchair. So, much to one of the young physicist's chagrin, only one of them got to cross the Atlantic in the lap of luxury with a box. It is public money that pays for this, after all.

The physicists' were not as concerned by the alpine terrain they would not be covering with their four-wheel drive baby carriage, as by the fragility of the irreplaceable device within it. For inside the state-of-the-art box there lay the Pixel Detector Endcap, "the most complex silicon tracking detector ever made," according to the physicist who headed the one hundred strong team that built two of a kind – one for each end.

The physicists handed out a short description of their 'precious baby' to the crew and the curious passengers, so as to quell any fears and do a bit of outreach by the same token. And since the Pixel Detector is the smallest in size, hyper sensitive and the closest to the interaction point, it will degrade the fastest and need to be replaced. The next generation is underway as you read this, so do not be surprised if you see a physicist, a pram and a box the next time you are lucky enough to travel First Class.

› *The Pixel Detector Endcap may be the smallest detector in size but it has the largest number of channels with the highest data output.*

ATLAS
BIG WHEELS

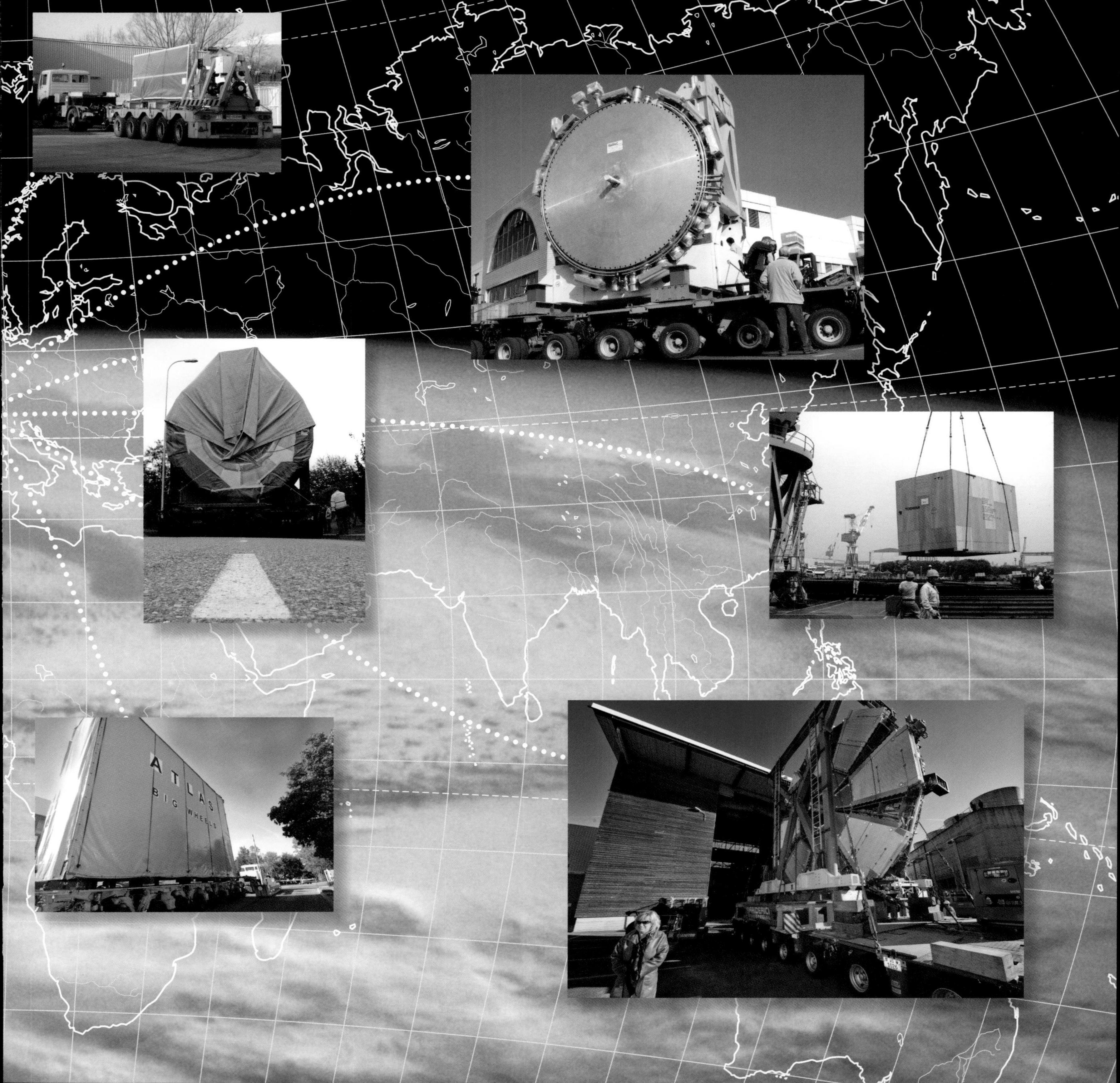

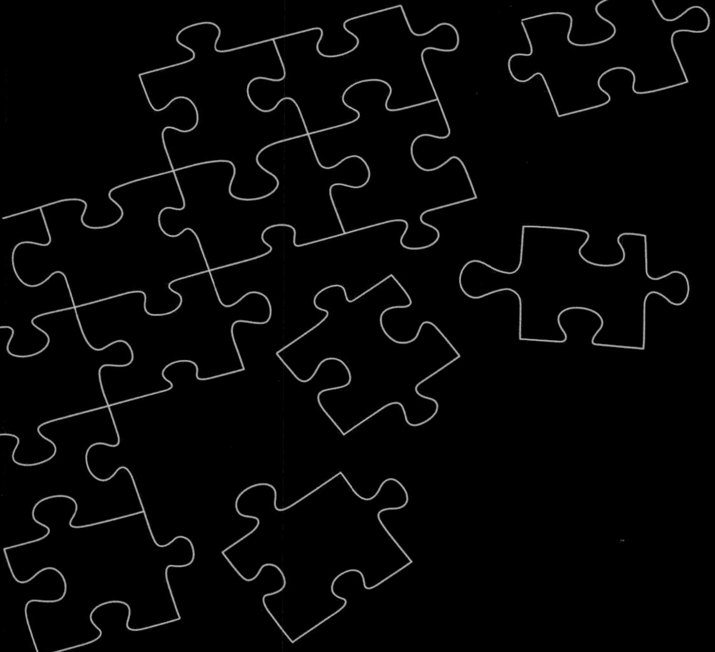

09

"ATLAS is like an onion, every layer involving a different technology"

Piecing the Puzzle Together

The tension at CERN was palpable when the first pieces of ATLAS started arriving in Meyrin and the building of the puzzle truly commenced. The air in some of the assembly halls was electric at times, with most of the work taking place on the surface. Phase one entailed checking that pieces fit together, phase two making sure that they worked together, and all this before they got anywhere near the cavern. Once painstakingly lowered into their underground home, the pieces were tested once again to ensure that no damage had occurred on their way down the shaft.

For the community as a whole, this lengthy period was tremendously important. Individual laboratories all over the world (as well as partner companies) that had built components of ATLAS now had to depend on other groups to help them complete the colossus. This was to be the first time the Collaboration really came together, moving from independence to interdependence, learning to trust and rely on the work and dedication of others, whilst being reliable partners themselves.

Orchestrating Test Beams

Picture this: ATLAS as an inordinately large onion, each layer a different sub-system with its very own technology. Eyes watering yet? They should be - in amazement at the sheer complexity of the beast. Over a decade, increasingly complex trials were run to verify the integration between the infrastructure and the technology. 'Test Beam' runs use particle beams over a range of energies to examine parts of the apparatus under realistic conditions.

Running a Combined Test Beam with multiple components is a bit like rehearsing an orchestra. You have to first convince everyone to come together to play; every musician must then practice playing alongside the others; and finally the full ensemble must function in harmony. Much like a symphony, timing is of the essence during these runs and any deterioration in the performance of one piece due to the presence of others must be evaluated immediately, so that the whole can be fine-tuned to work together.

Practice makes perfect

Orchestras rehearse for a series of concerts for months, sometimes years, to truly play as one. In much the same way, 200 scientists and technicians ran the Combined Test Beams over a six-month period - preceded by three years of preparation - to adapt the software, learn to perform together and collect long awaited data. Comprising 40 percent of the group, students played an important part during this period, with professors often foregoing the opportunity to participate in the tests to let them attend.

Student overtures

Many of these students had never had the opportunity to work on a large functioning detector. Most had worked on ATLAS for years, but running tests and taking data at this early stage was their first brush with a working experiment of this scale. They set up the tests, had to troubleshoot problems alongside more experienced physicists and engineers, took data and analysed it. Interacting with so many different people was also a tremendous experience for these students, many of whom earned their diplomas having at long last entered into the world of live detectors. Over the six months and approximately one thousand shifts, an exciting and friendly atmosphere developed. To this day if a problem arises with a particular instrument, the Test Beam participants know who to go to - the musician who plays that particular melody.

› *The ATLAS subsystem components, much like the instruments of an orchestra, need to rehearse together to achieve optimal results and work in harmony.*

› *Inner Detector*
Russian dolls: the Semi Conductor Tracker (SCT) is inserted into the Transition Radiation Tracker (TRT), in turn inserted into the Barrel Cryostat. The Pixels come at the very end.

› The Pixel Detector is the smallest of the Russian dolls; however, for technical reasons, it is the last to be inserted into the heart of the detector.

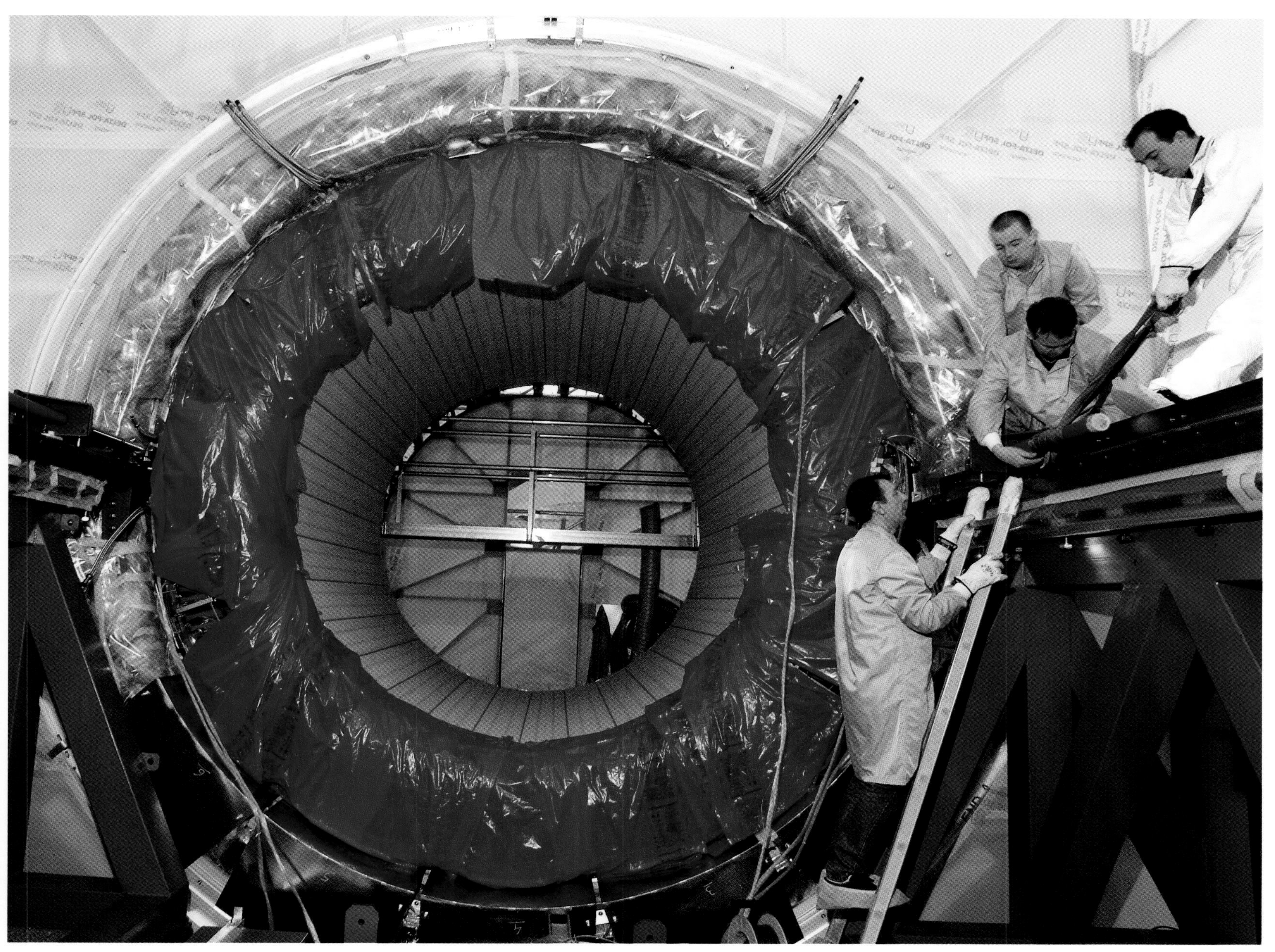

› *The Electromagnetic Calorimeter's first half-barrel is one of the earliest components to disappear into its cryostat in the heart of the detector.*

› *Liquid Argon Calorimeter*
600 kilometres of cables connect the calorimeters to the read out. Once they are installed in their cryostats they are no longer accessible.

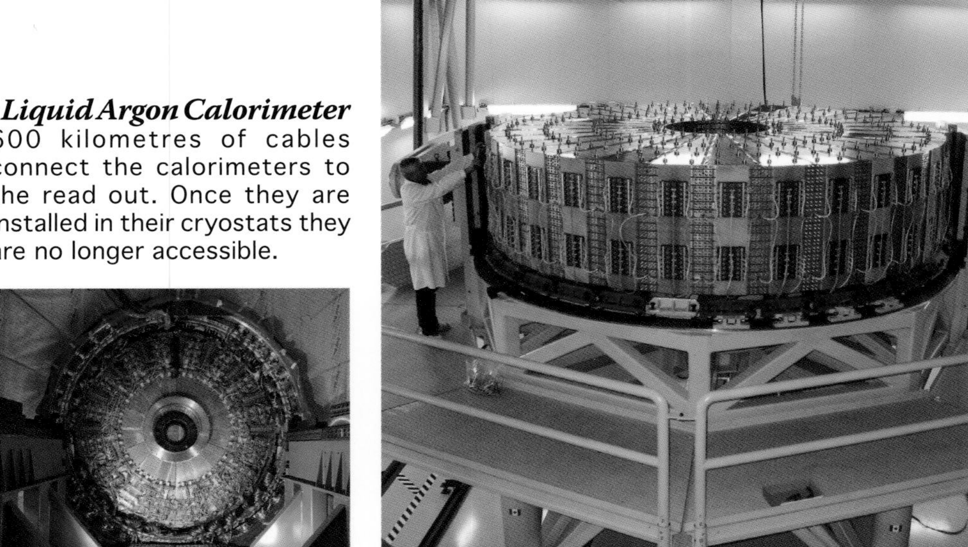

› ***Tile Calorimeter*** - Module after module is carefully stacked and surveyed to form a perfect eight-metre tall cylinder, each part placed with millimetre accuracy.

› ***Muons*** - The total surface of the muon chambers amounts to 20'000 m², roughly the size of three FIFA approved international football fields.

LOXAM

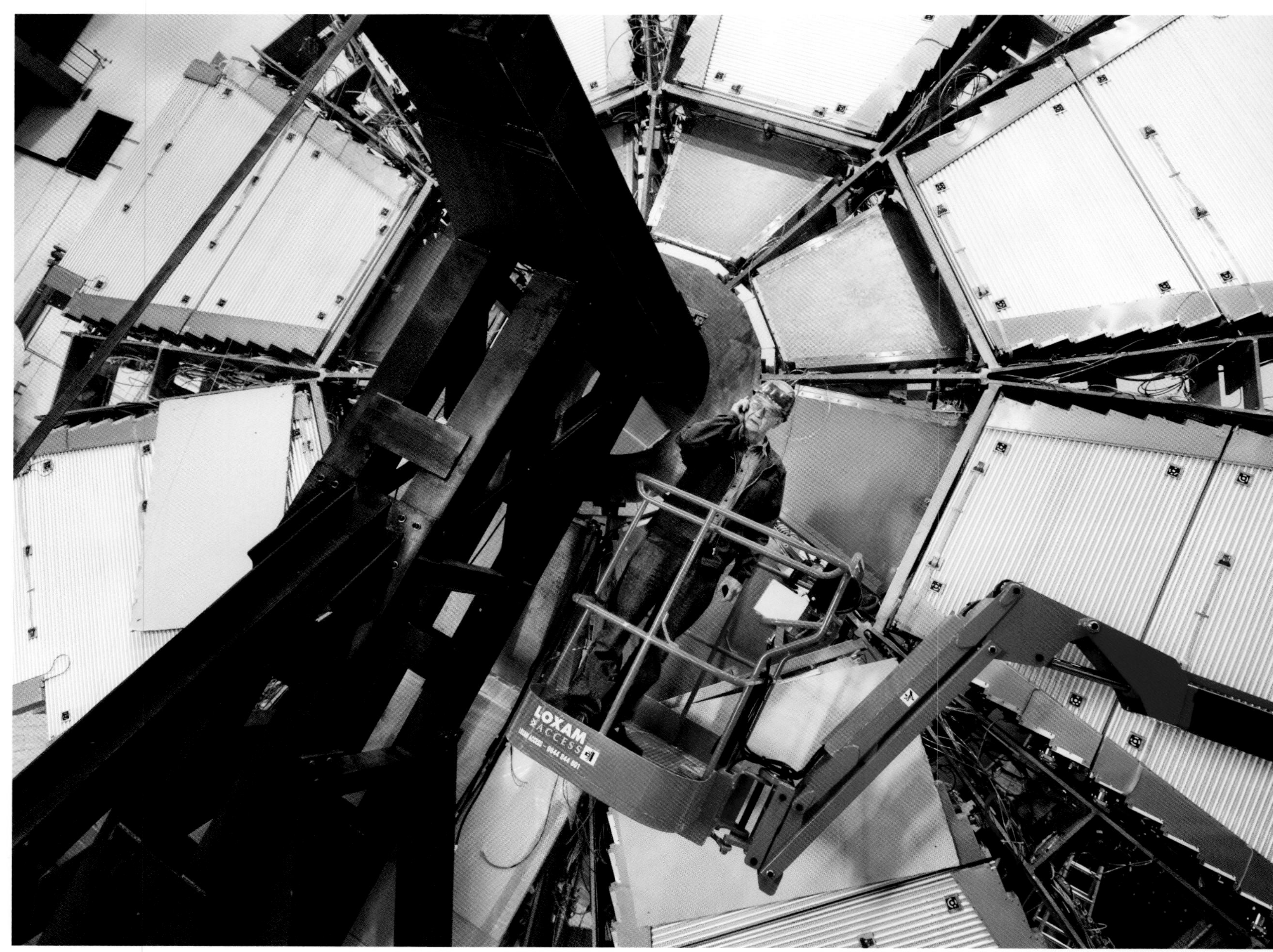

› *Each precision Muon chamber was shipped to CERN in a cradle to minimise vibrations during transport. No two mechanical engineers designed the same cradle.*

One of the few places in ATLAS with space to spare. These engineers in the Endcap Toroid will soon be replaced by immense magnetic fields.

› ***Magnets.*** The huge Barrel Toroid coils are assembled and tested on the CERN site before their short but epic journey across the road to the ATLAS cavern, while the superconducting Solenoid travelled all the way from Japan.

› ***Roman Pots*** - One of the last parts of ATLAS to be conceived, the Roman Pots rose out of a need to know precisely how much 'light' comes from the LHC.

› ***LUCID*** - Narrow gas-filled detectors form the Luminosity Monitor Detector (LUCID), which will make some of the earliest measurements of the LHC luminosity.

10

"Physics teachers shouldn't be afraid to teach 21st century physics"

Eye Care

Physicists are always thinking outside the box. They are forever pondering novel uses for their groundbreaking inventions and cutting edge measuring instruments.

To non-scientists, the world of physics often seems quite remote and somewhat out of touch with the 'real world'. Appreciation of the need for, or indeed use of, pure research, in any field, is often hard to come by. 'Selling' or 'justifying' high-energy physics is a constant in conversations beyond laboratory grounds and within funding institutes. Yet the contributions made by physics to everyday life are remarkable.

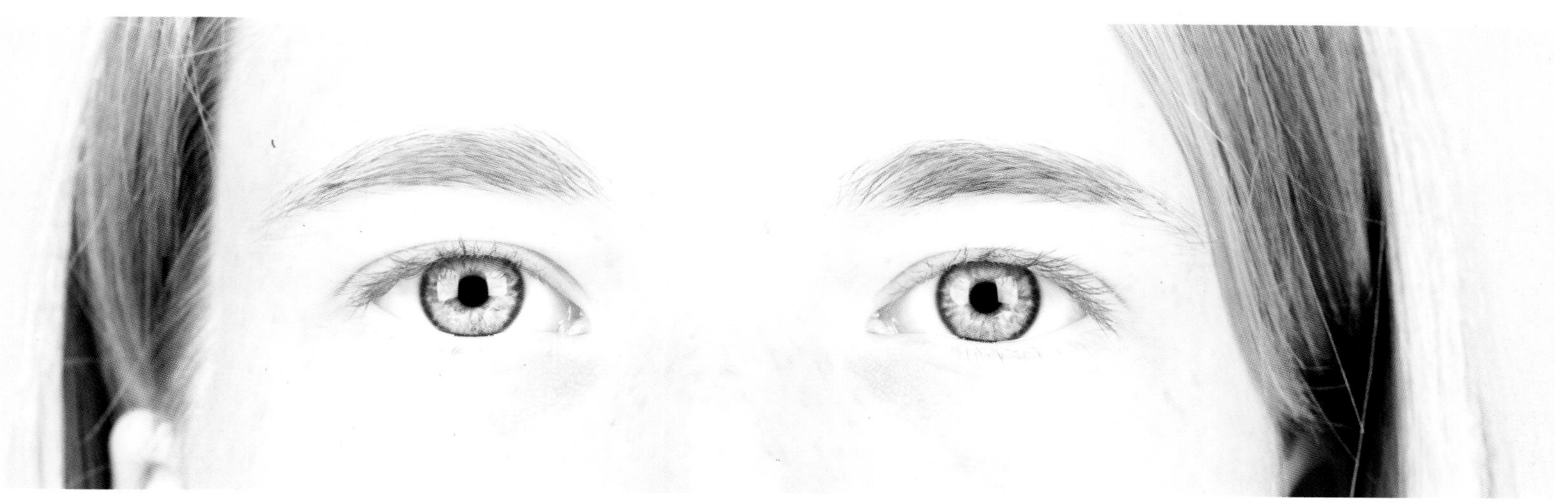

Think World Wide Web - CERN was the first to. Physicists also perfected the technology that led to radiotherapy and Positron Emission Tomography (PET); scanners, which use antimatter for medical imagery; particle accelerators, which are being continuously improved, can now target tumours with unprecedented precision. The technology transfers from particle physics to society at large have had a global impact, saving lives, connecting the planet and preserving our heritage.

Stay Tuned for IRENE

The link between high-energy physics and an American folk and blues musician may not seem obvious to most, but to certain lateral thinking physicists it is. What do we know about this link? Only that we have a match between the optical measuring techniques used in the Inner Detector and a musicologist called IRENE. Curious indeed.

So, who is IRENE? The fruit of a Russian physicist's romantic imagination and an American physicist's creative mind, IRENE personifies multiculturalism. She recently moved from a laboratory at Berkeley to the Library of Congress in Washington, D.C. and is currently undergoing tests to bring historic sound recordings to a large public. Her real name is Coordinate Measuring Machine. IRENE is her nickname (Image, Reconstruct, Erase, Noise, Etc.) and reproduction her passion - preferably to music.

'Goodnight IRENE'

IRENE's story begins at the turn of the century. A physicist involved in the ATLAS inner detector heard a radio programme about collections of sound recordings in American archives being at risk. At the time, precision optical image processing methods were being used in ATLAS to measure and align each of the 16'000 individual silicon detectors of the Semiconductor tracker (SCT). The physicists working with these techniques realised that they could be adapted to old recordings. By optically measuring how a needle moves over the shapely grooves of a mechanical disc, they could figure out the sound. Even broken or chipped discs can be mapped thanks to this digitised process, and scratches or dirt can be retouched on the digital image.

...And good luck

Three years and lots of experimentation later, the team wrote a paper on IRENE and sent it to various national bodies including the Library of Congress. They jumped at the opportunity, seeing value both in preservation (transferring analogue media to digital) and accessibility (digital being available to a large audience). A grant later, and IRENE was built. The first record the team digitized was 'Goodnight Irene'; an American folk song by Leadbelly. The name stuck - the former that is. She is now undergoing tests for a whole range of sound recordings, from 78 rpm shellac jazz records of the 1920s and '30s, to lacquer transcription discs of early radio broadcasts and vinyl records. A second version of the machine aimed at the wax and plastic cylinders first popularised by Thomas Edison is being fine-tuned at Berkeley. IRENE may well have a reproductive partner soon; in the meantime, she is definitely the grooviest time machine around. Stay tuned ATLAS.

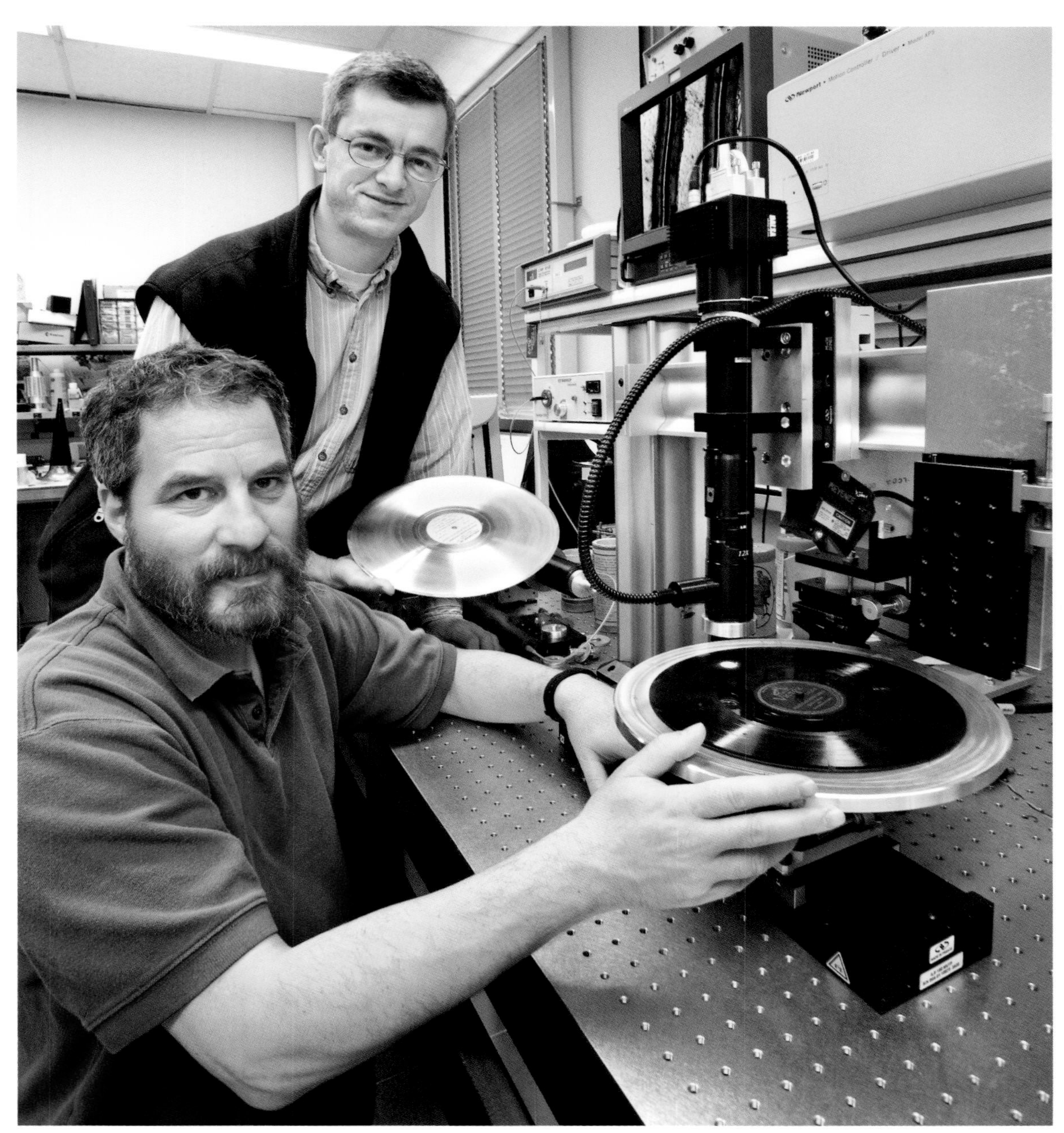

› *The Optical Sound Recovery System, codename IRENE, is contributing to the preservation of our recorded heritage for current and future generations.*

Making Light of Dark Matter

The Outreach Group at ATLAS has a very carefully honed strategy to spread the word on today's physics and attract young minds. The brochures, films and talks serve as bait to entice students and others into the fold of modern physics. Gone are the days of teaching 18th century physics to 21st century students. Now is the time to introduce quarks, dark matter and supersymmetry into the classroom.

Overall, the sciences are attracting fewer students, yet the advances in this field over the past decade have been prodigious. Making the sciences more alluring to both teachers and students is a necessity and particle physics certainly has the tools and the credentials to do so. Winning over high school pupils is one side of the equation; on the other, teachers must be kept abreast of the state of the art. The Outreach Group has placed the ATLAS experiment on the education frontline and is participating in several international education projects.

Mastering the Universe

Among these projects are the international Masterclasses offering fascinating opportunities to inquisitive young minds. The classes enable senior pupils to access real data from a recent experiment, learn how to identify and work their way through various types of interactions visualised by the analysis software, and contribute

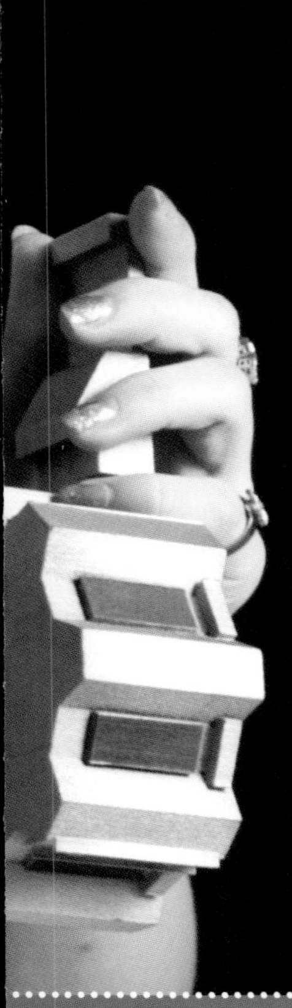

› *A professor visits a high school class to discuss current developments in particle physics and further stimulate interest in the science.*

collectively to the measurements. Masterclasses offer the opportunity to experience the taste and excitement of 'real' physics and to interact with enthusiastic professional physicists. In order to keep up with their students' newly acquired knowledge, the teachers must also brush up on theirs.

Students Beware

The ATLAS Outreach Programmes aim to approach physics in a creative manner in order to capture people's imaginations. Episodes I and II of the 'Star Wars-like' ATLAS film do just that, opening the audience's eyes to the scope and scale of particle physics, the remaining unknowns food for ambitious young brains. It is no longer about apples falling from trees or where to put that darned decimal point, but about dark energy, the exploration of the unknown and what happened less than a billionth of a second after the Big Bang.

Physics has never been more fascinating: students beware, ATLAS will make particle physicists of you yet.

Eyes Wide Shut

There are no national or international bodies overseeing the sciences and brainstorming for possible applications between the sciences and other fields. When a crossover takes place it is more often than not a group of scientists who come up with an idea, discuss it in their spare time, run a few trials, write a paper and approach particular companies, institutes or individuals. Overall, it is very ad hoc and only happens if people are truly driven.

This is exactly how IRENE came about. A similar connection has been made between biology and physics. "If you have a background in physics and come into contact with neurobiology, you can do amazing things," says the physicist behind the Retina Project.

With an eye on the future

He was working on silicon microstrip technology at the Santa Cruz Institute for Particle Physics, when he came into contact with biologists working on the retina. Soon he realised that the very precise instruments developed for the detector could help discover the secret language used by the eye to send information about the outside world to the brain.

By understanding how the retina works, the way in which it is wired, the electrical signals it sends, the encoded images transmitted to the brain and how the information is processed, scientists could eventually feed information into the retina to create artificial sight. Retinal prosthesis implants are still at the experimental stage, but have already helped some blind individuals. This, in and of itself, is quite miraculous. Thanks to this collaboration between physicists and biologists, future retinal prosthetic devices may be better able to restore normal visual function to people who have never before seen the world in all its glory.

› *Detail of a retinal ganglion cell. By studying the way the retina is wired scientists learn more about how the brain stores and recalls information.*

Sleep tight

In the world of anaesthesia, physicists have successfully adapted tools that allow for more precision in the measuring of gas ratios. The ultrasonic technique uses the speed of sound to analyse gas mixtures, with sound passing more slowly through heavier gases. Software developed to measure the concentration of refrigerant gases in particle detectors has also been adapted for use in oil refineries and in the semiconductor manufacturing industry. Unfortunately, the physicists and technicians from CERN, the University of Melbourne and the Marseille Centre for Particle Physics (CPPM) who ran convincing trials at the Hôpital

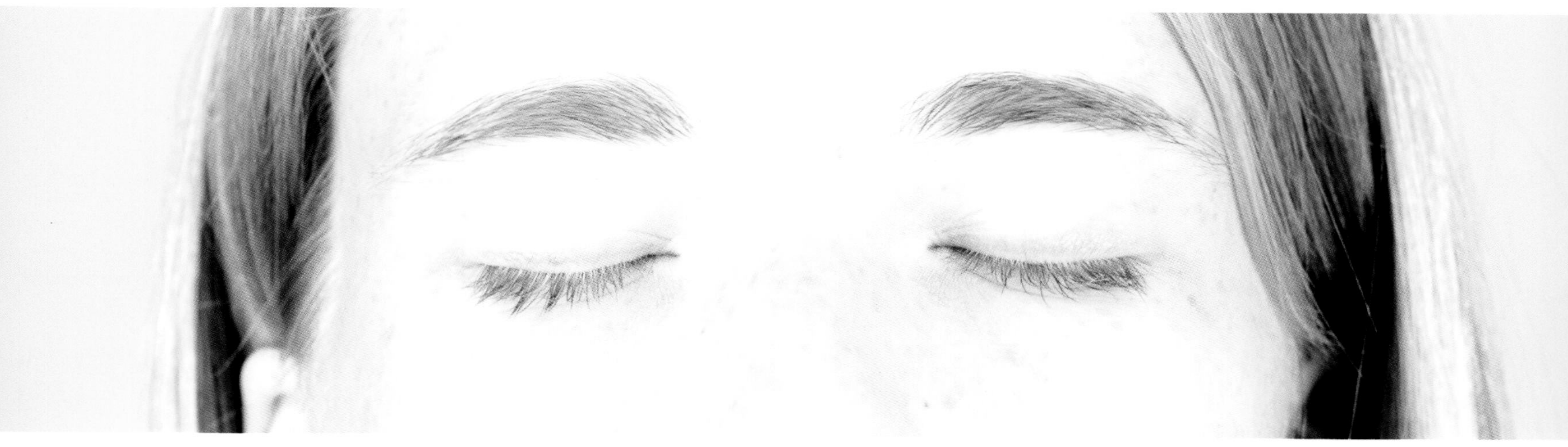

Cantonal de Genève, did not have their gas analyser taken up. Two out of these three transfers were adopted though, which just goes to show that when physicists put their mind to something, extraordinary possibilities open up.

11

"The action is over lunch and coffee, not in the sessions"

Together Forever

Imagine having to make well over 2000 scientists, engineers and technicians from 170 institutions in 37 countries feel part of a whole. This is a feat ATLAS has excelled at over time - two decades in fact. It is inevitable that people lose sight of the big picture when they are working on a very small part of such a large enterprise. The ATLAS Overview Weeks and the Trigger & Physics Weeks remind everyone of the breadth of the experiment, the importance of each and every aspect of it and of everyone's contribution.

Trigger & Physics Weeks are all about discussing the science the physicists want to study, the measurements to be made and how to go about them. Maximising sensitivity (of the detector, not the scientists) is at the core of these weeklong meetings when the various sub-groups come together. The ATLAS Overview Weeks cover the entire experiment, while the week abroad every summer gives participants the opportunity to interact more closely and attend the numerous social events after the sessions.

Spirited Meetings

Building and sustaining a team spirit across continents, cultures and a passionate yet diverse group of people is quite something, and ATLAS has pulled it off. The various types of Collaboration meetings, with the social events that accompany them, have brought the collaborators together, fostering a spirit like no other and a shared enthusiasm rarely encountered in the world at large.

Scottish country dancing following a generous whisky tasting in Glasgow; a nine-course meal in a medieval setting in Clermont-Ferrand; classical music in Prague in central Europe's oldest university; other-worldly singing at a church service near Dubna; an excursion to the Black Forest on the outskirts of Freiburg; an odorous cheese and wine tasting in a low-vaulted cellar under Paris; literary jewels in the relaxing surroundings of Brookhaven; the Lord Mayor's reception in the Blue Hall in Stockholm's City Hall where Nobel laureates are 'made' and ATLAS physicists look dreamy-eyed. These are some of the experiences that have made the Collaboration what it is today: strong, unified and vibrant.

Cross fertilisation

The ATLAS Overview Week abroad is quite the place to get a handle on the big picture and to meet colleagues from other institutes in an informal atmosphere. It is not so much the sessions that prove to be the focal point, but the chats over lunches and dinners when everyone is thrown together and exchanges are plentiful. Things happen in passing: students get to talk about their research with veteran kindred spirits; teams that are about to merge get to know each other better; physicists get answers from colleagues over coffee. In other words, work is done in real time.

Much the same can be said of the ATLAS Overview Weeks at CERN. They are just as important and informative; but with more distractions at hand - such as the detector itself - they can be slightly less productive in terms of personal interaction. One physicist describes these weeks with a grin: "It's a meeting where the people who are doing the work explain what they are doing to the people who aren't doing the work."

The perfect blend

Slowly but surely, the ATLAS Overview Weeks are blending into the Trigger & Physics Weeks with the experiment approaching its long awaited turn-on. These meeting weeks go into the nitty-gritty and involve the various physics working groups with rather enigmatic names. Each group looks into a particular aspect of the physics – with, for example, the Monte Carlo group playing a game of simulation, the Exotics (and we're not talking Copacabana here) looking for predicted but as-of-yet unseen particles, and the Top physicists waiting to encounter extreme quarks. Whichever week it may be, there is always time for coffee and the meeting of minds.

12

"Every single millimetre of space was fought over"

Tunnel Vision

"When the cavern was empty it was mind bending, so huge. Now that it's full of ATLAS it looks relatively small." Workers excavated 300'000 tonnes of rock 100 metres below ground to make room for a cavern the size of half a football field and the height of a 10-storey building. It took 400'000 hours of work to complete the colossal home.

The scale of the experiment takes on another facet when the infrastructure surrounding the detector is considered. The cavern alone represents a first in the history of civil engineering with a 10'000 tonne ceiling - equivalent to twenty times the 'Rainbow Warrior' - temporarily secured by 38 steel cables anchored to galleries 25 metres above before being released onto the walls.

"It was like building a ship in a bottle." Once the cavern was built ATLAS had to be lowered down bit by bit through the vertical shaft, much like a bottleneck, into the vast area below where cranes and cherry pickers looked like children's toys from the surface. And then the real fun began with thousands of kilometres of cabling, a host of safety issues and confined spaces like nowhere else.

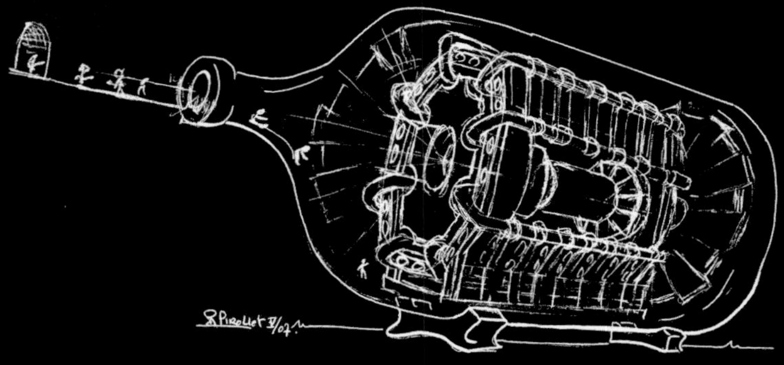

› *The excavation of the cavern, also known as 'Point 1' on the Large Hadron Collider (LHC), begins in February 2000.*

› *The integration of the detector in situ requires an engineering team capable of dealing with every detail of the installation. Major hardware components are manipulated and installed with millimetre precision thanks to studies of support and access structures, routing of cables and processes for lowering tools.*

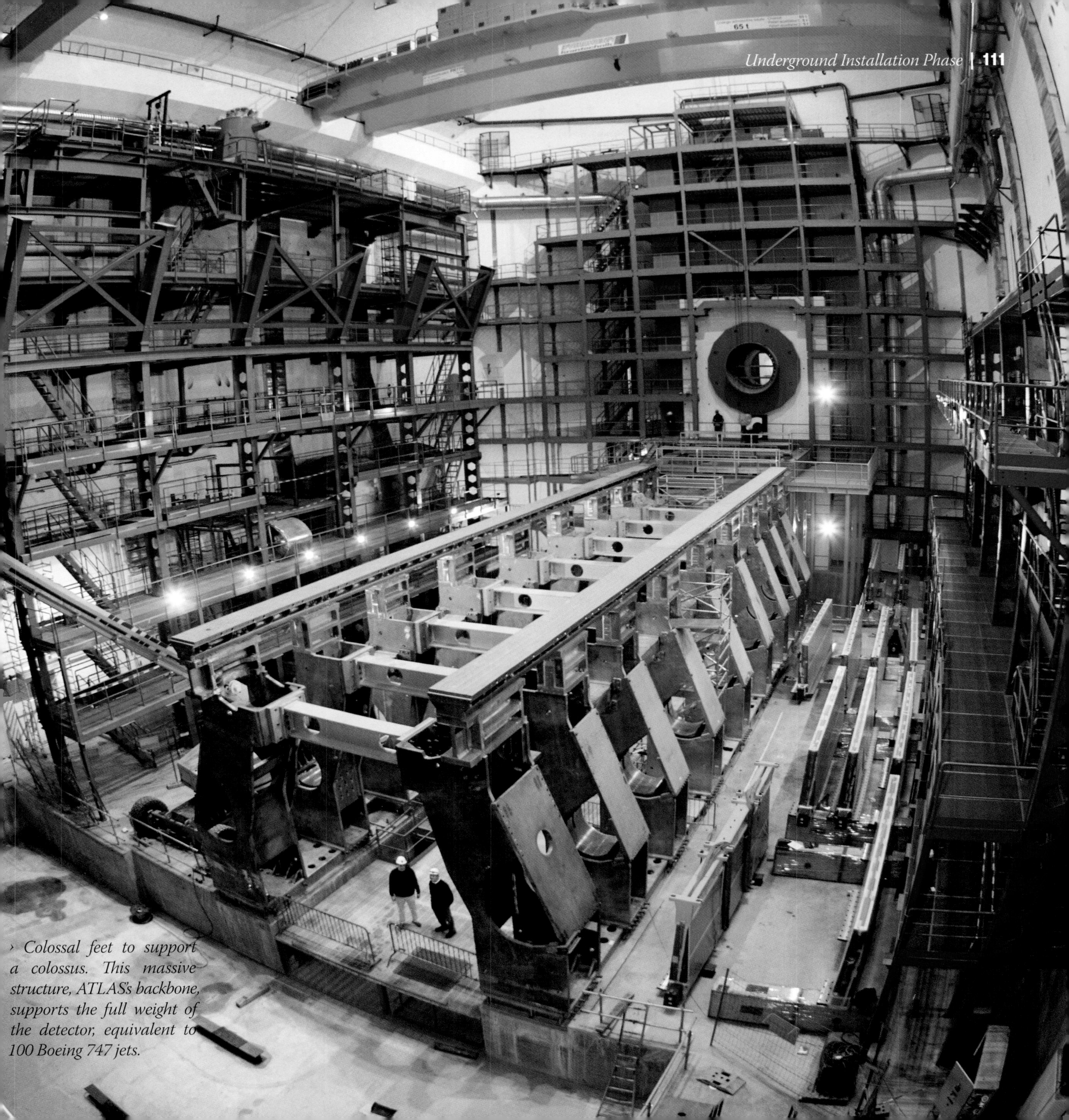

› Colossal feet to support a colossus. This massive structure, ATLAS's backbone, supports the full weight of the detector, equivalent to 100 Boeing 747 jets.

Nerves of Steel

Only a lucky few were allowed into the surface building each time to watch the gargantuan task of lowering the Barrel Toroid magnets into their new home. Adrenalin ran high on these days, and although there was no reason to whisper, everyone did - perhaps in awe of the crane driver's extraordinary skill and patience or for fear of disturbing the magnets with too loud a voice.

Four wheels were placed on the four corners of the Barrel Toroid coil to ease its trip down the shaft - wheelbarrow wheels, nonetheless. The eight 25-metre long magnets (nine if you count the practice run with a 'dummy'), were the most nerve racking and technically difficult parts of ATLAS to lower.

From an engineering point of view the challenge was immense, with only a few millimetres of clearance between the one hundred tonne magnet and the wall. Oh! And a rotation to perform at the foot of the shaft. Five hours of anxiety, nine times, over eight months, and not once did the professional crane drivers hesitate or lose their nerve. Number 7 did get a bit of a bash, but all in all everything went nail-bitingly smoothly.

› *One of the two Muon Small Wheels is lowered into the cavern. It is one of the last pieces to fit into the three-dimensional puzzle that is ATLAS.*

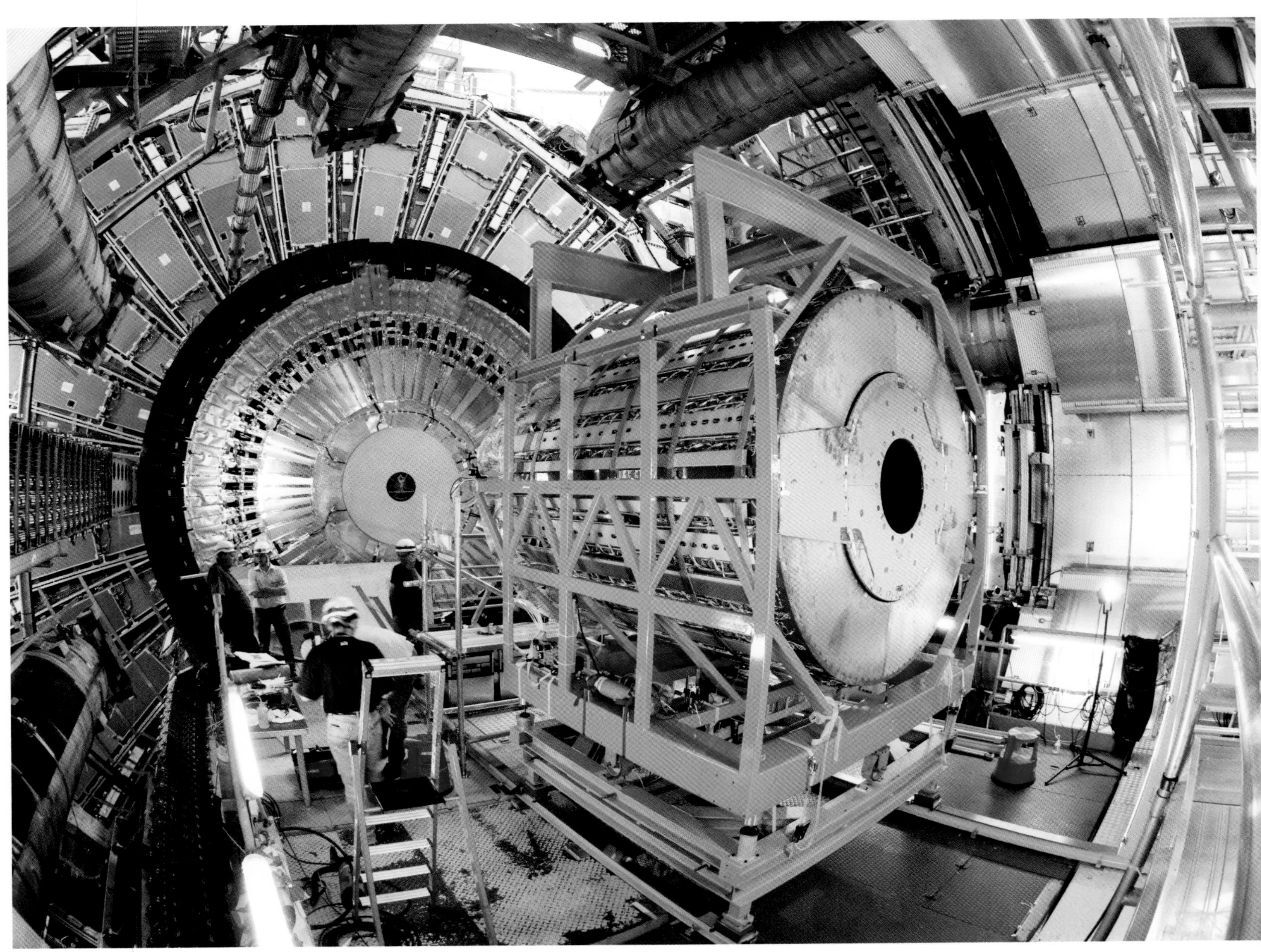

› *The Inner Detector Endcap C is moved into place. This 2.5m long by 2m diameter structure was installed with just millimetres of clearance.*

› *Inserting the nearly 7-metre long Pixel Detector package into its new home, The Barrel Cryostat.*

› *The ATLAS cavern from the underground up. The Endcap Toroid Magnet is in place after an epic, if ponderous, 3-hour trip down the shaft.*

› *Much like an arch, the Endcap Calorimeter modules were built on a cradle, the top half in a distorted cylindrical shape released with the last module.*

› *Each of the six Big Muon Wheels is composed of 16 segments, lowered individually into the cavern and then assembled together to form a circle.*

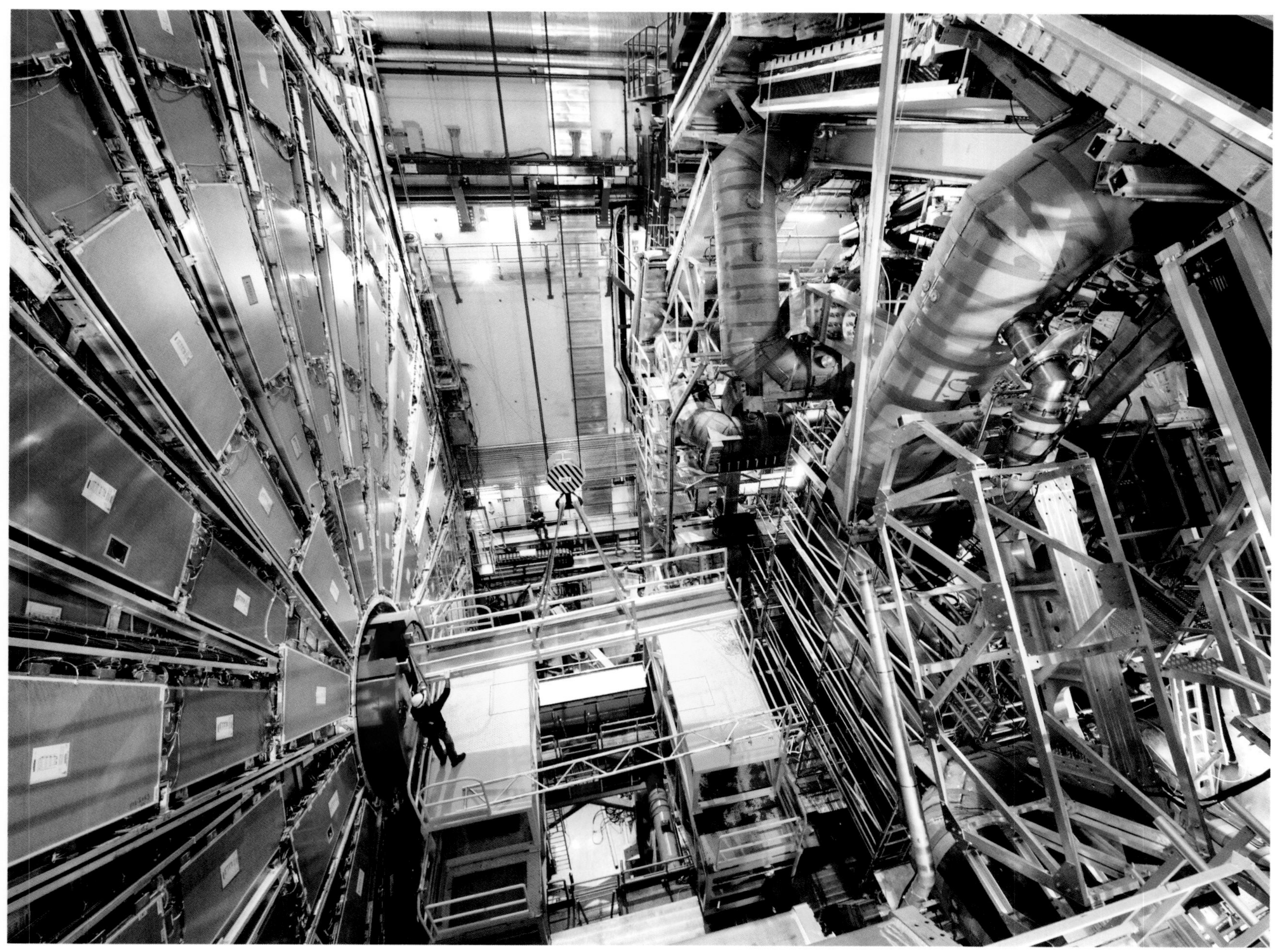

› *Professional crane drivers performed some of the most nerve-racking work with often only a few millimetres to manoeuvre pieces into place.*

› The first Inner Detector Endcap after complete insertion within the Liquid Argon Cryostat.

› *Before the cavern was fully excavated the LEP accelerator was still running. To save time, engineers decided to cast the cavern ceiling first, temporarily suspending it with steel cables to create the largest suspended vault in the world.*

› *Poetry in motion: the endcap calorimeter floats towards its final position on orange air pads. The choreography is flawless.*

› *The vacuum in the 27km beam pipe that runs through ATLAS has an atmospheric pressure ten times lower than the atmosphere on the moon.*

› *A 'cherry picker' accesses one of the two Endcap Toroid Magnets on the left, with the Muon Big Wheel on the right.*

› *Even the tiniest of objects falling from a height could perforate the MDT's thin aluminium tubes, hence the protective covering.*

Sealing Envelopes

When a survey measuring the exact size of the cavern revealed that it was 50 millimetres larger than planned, everyone was ecstatic. Five centimetres may seem like nothing in a cavern measuring 55 metres in length, 32-wide and 35-high but it is immense in an experiment of this kind.

Every millimetre counts when building a detector of this size in a confined space, and that means every single millimetre. The blueprint containing each and every bit of ATLAS is so precise that dimensions are determined down to 10 microns, approximately one-tenth the diameter of a human hair.

Stealing away

Each system and its subsystems have a defined space called an envelope, which must be respected. Some have negotiated over years for an extra millimetre or two. Meetings have taken place all over the world to solve the many issues surrounding space and find the right solution to each problem. With over 50'000 drawings, documents and 3D models in specially created databases at ATLAS, there is no cheating. It would take years to simply run through these massive databases to delete the obsolete versions.

Fortunately, between 5 and 20 millimetres were left between each subsystem just in case. Never before have a few millimetres been so tirelessly fought over.

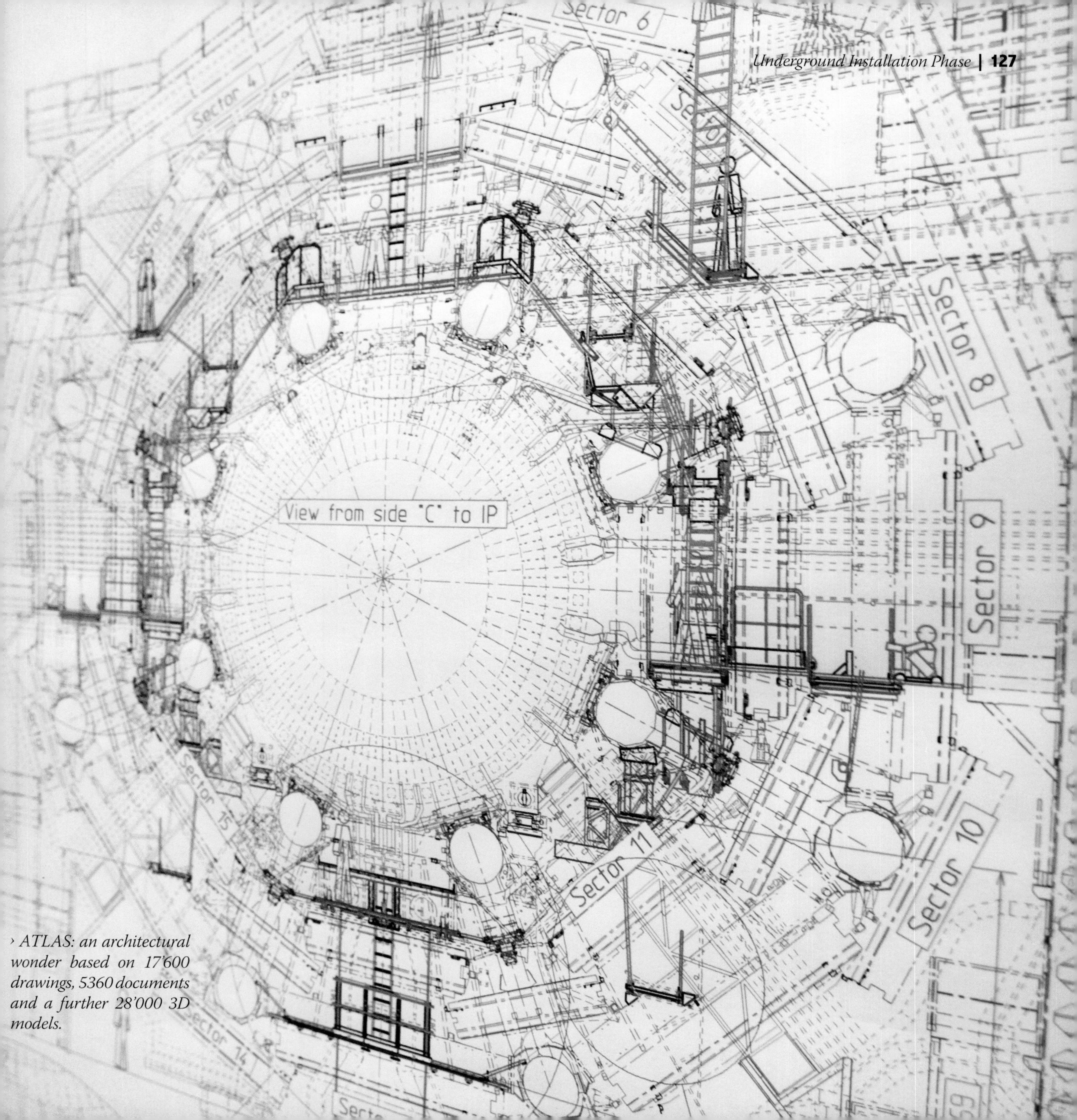

› *ATLAS: an architectural wonder based on 17'600 drawings, 5360 documents and a further 28'000 3D models.*

› ***Inner Detector*** - Months were spent connecting the Inner Detector's delicate optical fibres, cables and pipes. One last inspection and the way is clear for its installation in the cryostat.

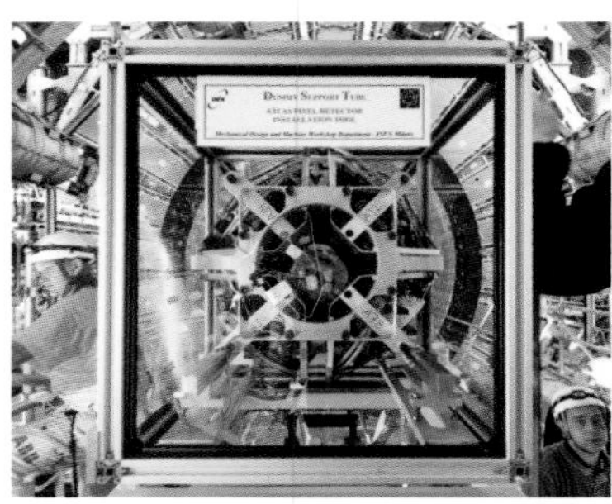

› *Liquid Argon Calorimeter*
The three cryostats (the Barrel and two Endcaps) that house the Liquid Argon Calorimeter hang by a thread as they descend.

› ***Tile Calorimeter*** - The Tilecal converts the energy of hadrons into electrical signals via layers of lead, scintillator tiles, optical fibres and phototubes.

› ***Muons*** - Physicists will be able to reconstruct muon tracks in this behemoth of a detector with 100-micron precision (roughly the thickness of a human hair).

› ***Magnets*** **-** The awe-inspiring Superconducting Barrel and Endcap Toroid magnets permit precise momentum measurements of charged particles.

Tying Up Loose Ends

Layer upon layer upon layer - it is quite overwhelming to see just how many there are. We are talking cables here, thousands of them - tens of thousands in fact - all labelled, all different colours and sizes, everywhere, in front of you, more hidden underfoot and yet more overhead. Unfathomable.

ATLAS's blood vessels and nervous system are the cables reaching deep inside it, bringing lifeblood to the machine and feeding vital information to the brain, the control room beyond. At its heart lies the Inner Detector; with one-third of all cabling, it has the densest concentration of blood vessels in the experiment.

Wired

The sheer amount of cabling brings home the complexity of the machine. "We have to connect this thing, otherwise we can't use it," a physicist explains wryly. And connect it they have. There are cables bringing power in and cables taking information out. Most of the 50'000 cables, give or take 5000, either start or end in one of the two counting rooms spanning three levels. For two and a half years teams from Russia, the Czech Republic and Poland installed 3000 kilometres of cables for the Inner Detector, and with some measuring over 100 metres long and needing to be carefully placed tens of metres overhead, they had a delicate task indeed.

› *3000 kilometres of cabling course through the detector.*

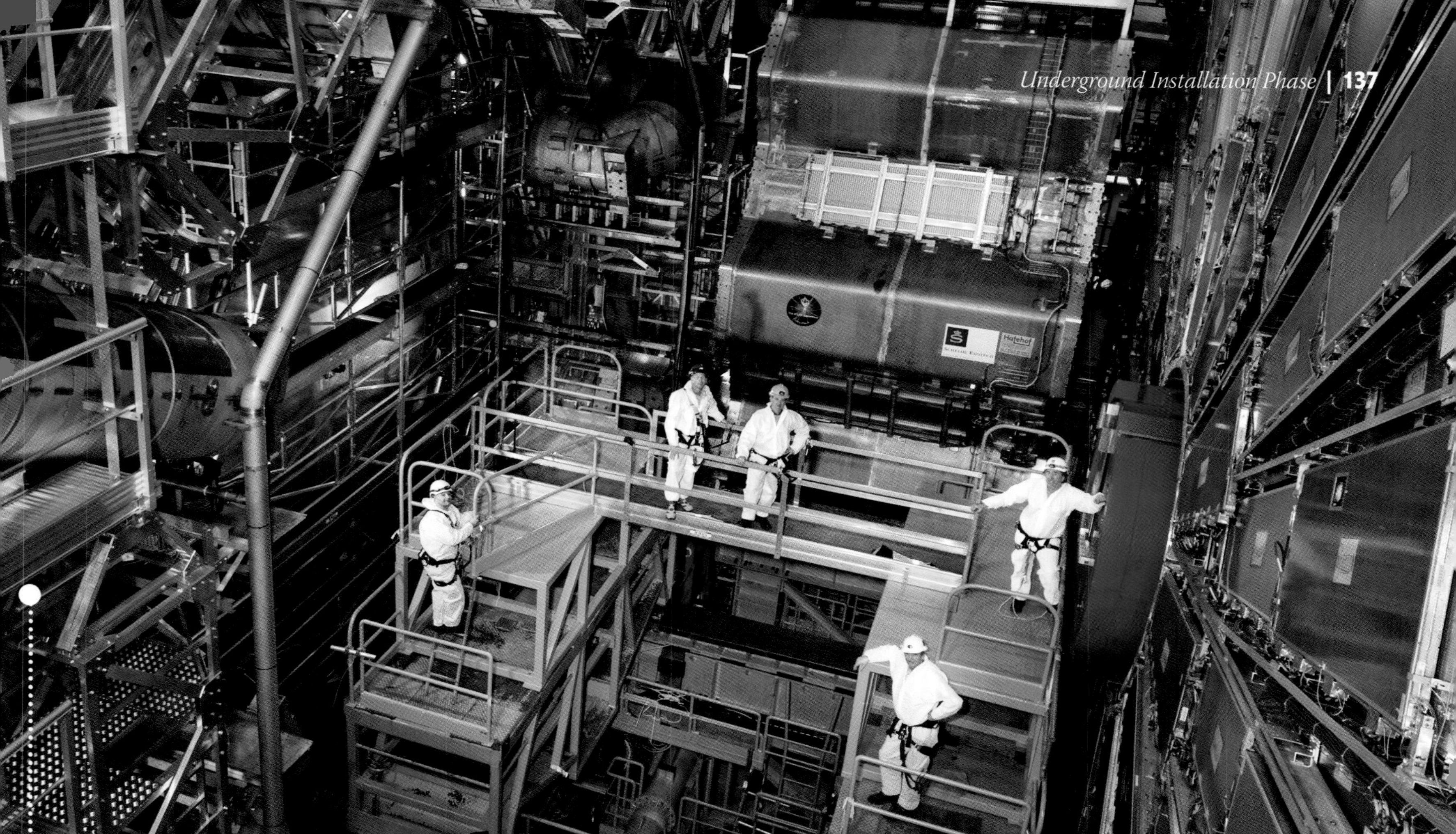

Harnessing Safety

Job Opening: Seeking brilliant physicists to work at ATLAS. Must be skilled at mountain climbing, have no fear of heights and be able to fit into small spaces.

Although this advert has never run, it is closer to the truth than you may think. One of the many safety courses on offer at CERN is all about harnesses and climbing vertical shafts. Experienced mountaineers teach sometimes nervous physicists the art of scaling walls and large infrastructures such as detectors.

Safety is a big deal in an environment such as ATLAS. Numerous safeguards have been put into place to avoid just about every possible emergency. To work in the cavern, the engineers, technicians and physicists must take various safety courses and wear a hard hat and steel-toed boots. And once the detector is fully functioning and exposed to radiation, security will increase tenfold. The priorities in case of emergency remain the same, though: preserving lives first, then the environment and lastly the equipment.

Behind these many layers of security is a team comprising safety officers for cryogenics, radiation, flammable gases, laser beams and electrical appliances – and we're not talking toasters and kettles here – territorial safety officers, and a host of others with equally impressive titles. But the best is yet to come, so keep your eyes peeled for the futuristic retina-scanning devices that will spring into action when ATLAS does.

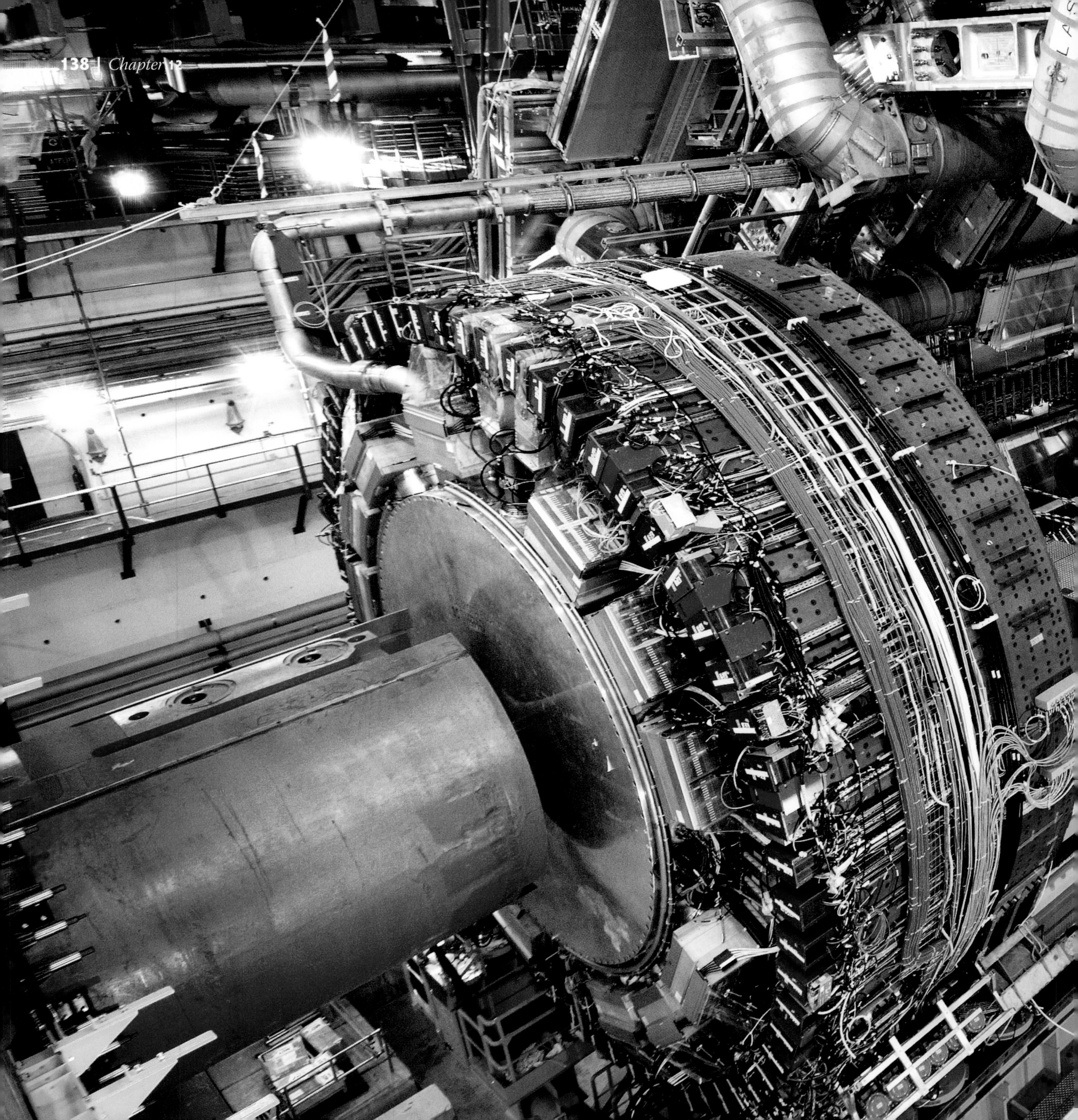

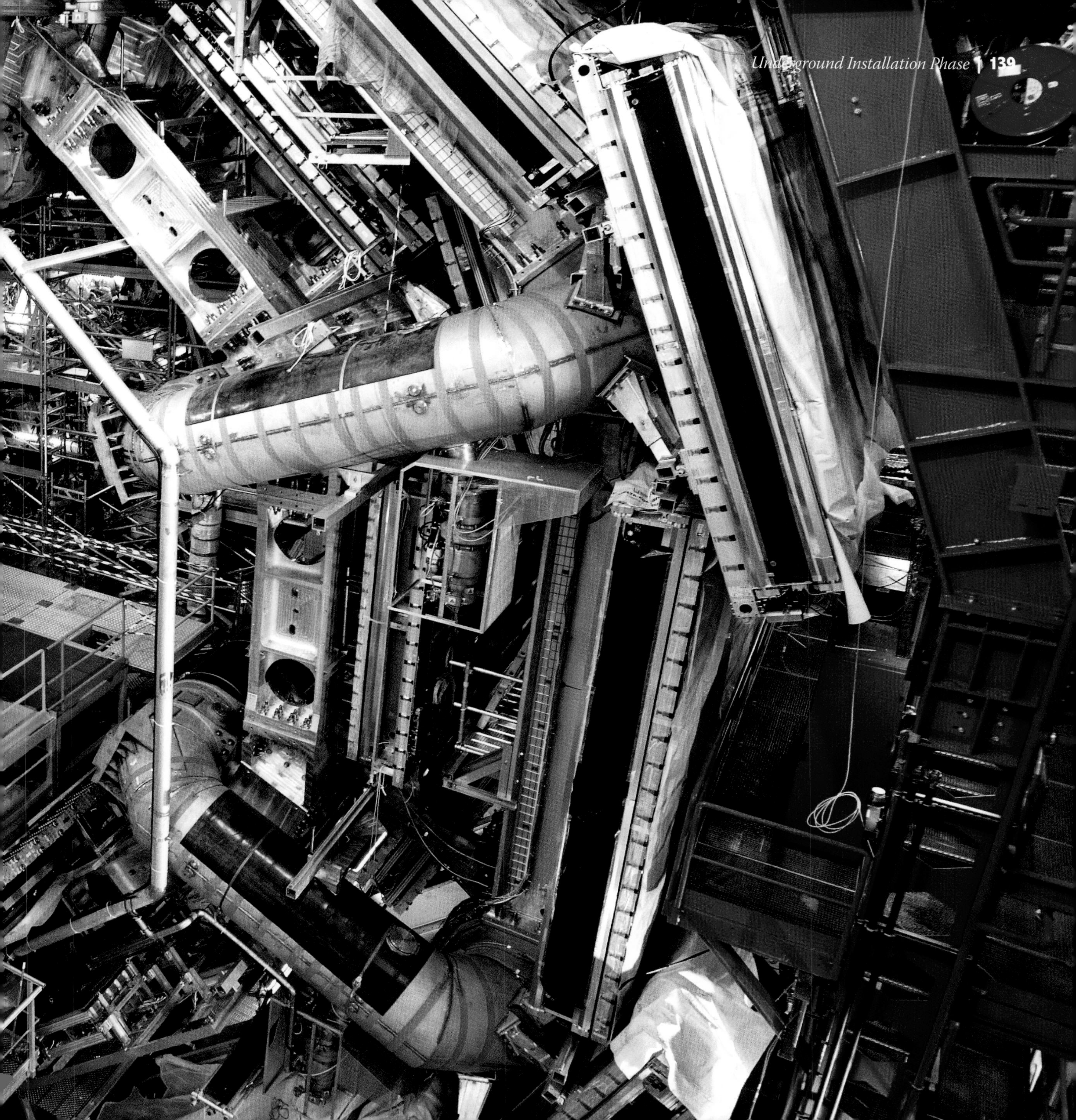

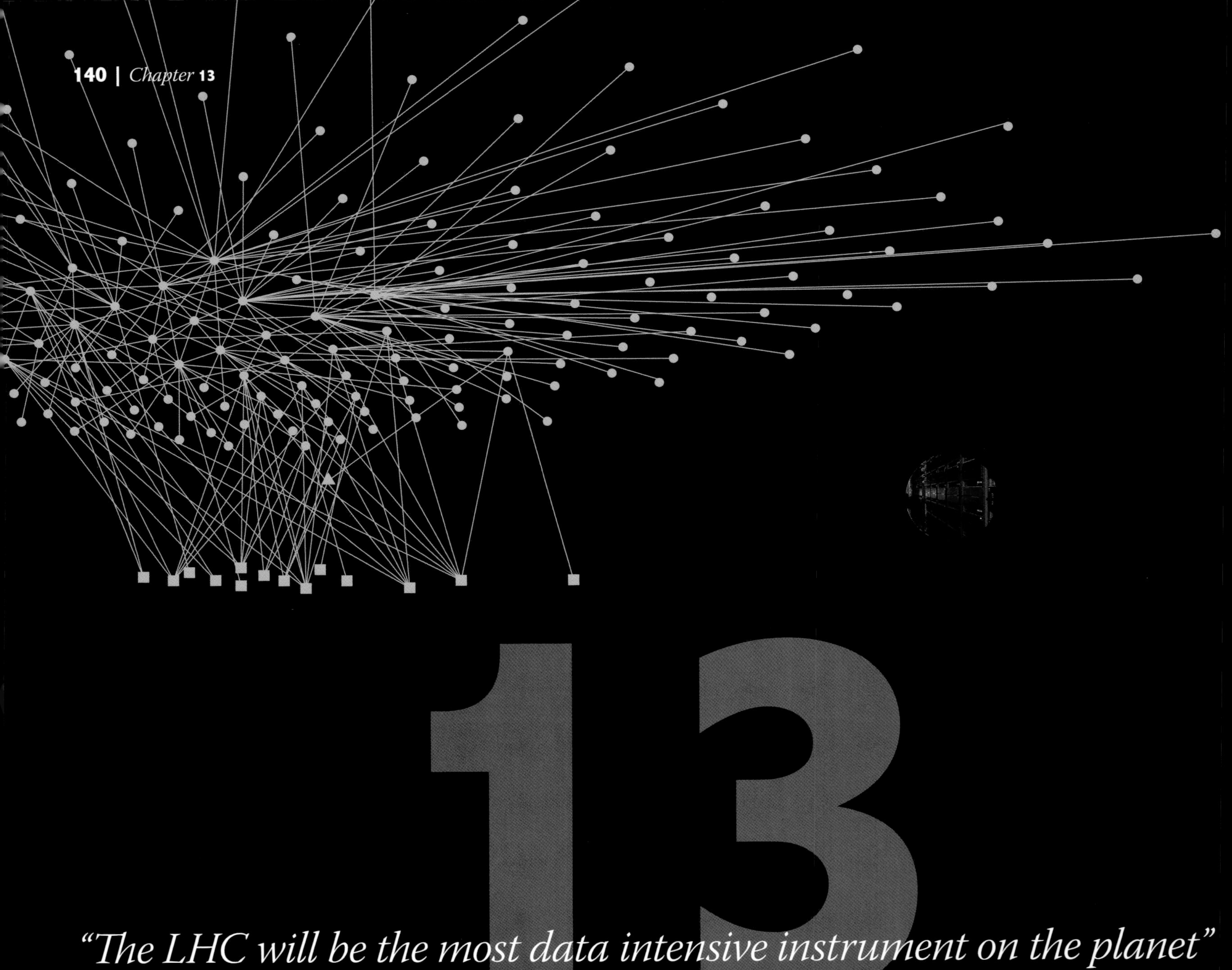

13

"The LHC will be the most data intensive instrument on the planet"

Grid[un]lock

Things are stirring at CERN. Change is in the air. Could a new revolution be brewing? CERN is where the World Wide Web was born, after all, so the rumblings must be well founded. Could this forthcoming revolution in data storage and computer power have anything to do with the Grid?

CERN and ATLAS certainly seem to think so and are spearheading some of the most exciting and far-reaching developments in Grid technology. Whereas the Web is a service for sharing information over the Internet, the Grid is a service for sharing computer power and data storage capacity over the Internet. Ultimately, its aim is to turn a global network into one vast computational resource.

This worldwide Grid will store and analyse the torrent of information that will gush forth once particles are let loose in the LHC. One hundred thousand of today's most powerful personal computers would be needed on site at CERN to do this job if this next step were not being developed. It certainly sounds as if the Grid has the makings to become quite the revolutionary type.

› The foundation of the data Grid, the robotised magnetic tape vault, constitutes Tier 0. The tried and tested magnetic tape, a robust and mature technology, is used to store the complete LHC data set, from which a fraction of the data is copied to overlying disk caches for fast and widespread access.

Two For All

ATLAS hopes to be the bearer of a host of unexpected discoveries. Finding the Higgs boson will be the crowning achievement of the Collaboration. This long-awaited and groundbreaking physics will first reach us through two rather insignificant looking yellow cables feeding gargantuan amounts of information to racks of computers and the Grid beyond.

The data generated by the LHC and its four experiments will be vast. So vast, in fact, that it will equal one percent of all information humanity produces each year – including books, digital images, music and movies. These 15 Petabytes are equivalent to 15 million Gigabytes, or enough data to fill about 100'000 MP3 players, every year.

From homemade to globally-made

We have come a long way since the seventies when particle tracks were recorded on tens of thousands of photographic negatives, and 'scanning girls' looked at each one, tapping into computers the beginning and end coordinates of particle trajectories. Forty years later, hundreds of thousands of computers the world over will work together as a single, huge and powerful unit to deal with the onslaught of data. The challenge in this day and age is not only storing and analysing the data, but also making it readily available to thousands of scientists around the world.

Up until now it has been difficult, expensive and sometimes simply impossible to achieve certain scientific goals with current information technology. Increasingly complicated problems demand ever-increasing computer power and data storage. Just think of the sequencing of the human genome and its three billion chemical units; or of earth scientists downloading from space hundreds of millions of images daily. The Grid will make data quickly accessible, so that computation can take place where the data is.

RC

Efficient carpooling

Unlike the Web, there is not one single Grid: some are private; others public, regional or global; and others still are dedicated to one particular scientific problem. With the Grid the user will not have to worry about where the computing resources come from, much like switching on a light and tapping into the electrical power grid. Ever the innovator, CERN has teamed up with a host of European research centres to develop a European Grid infrastructure for all sciences.

The philosophy behind CERN's Grid is efficient resource sharing on a global scale. Much like a car pool, many different people own Grid resources. Sometimes you share your car with others, and at times they share theirs. You do not necessarily know these people, but since they are in the same car pool you trust them, and if someone breaches that trust then they must bear the consequences. The organisations that make up the Grid must also abide by certain rules, create radically new security mechanisms to foil hackers and find ways to regulate access to valuable computing resources.

Tier-theory

So how does the Grid work in practice? All the data emerging from the various detectors is stored at the CERN computer centre, known as Tier 0. The network then sends the data to Tier 1, consisting of 11 computer centres in Europe, Asia and North America. They receive the data from the experiments they are interested in and store it to tape, providing multiple backups of the raw data. Approximately 150 universities and computer centres form Tier 2. To access this common infrastructure, wherever it may be, a person must be registered with an experiment or university and have a Grid certificate. By spanning the globe, the Grid will bring science within the reach of both the richest and poorest institutes everywhere.

› Between three and five Gigabytes of data per second - 200 events filtering down from the Trigger/DAQ - will flow from ATLAS through a single yellow cable to Tier 0. The Grid will then kick in, sharing computer power and data storage capacity over the Internet.

14

"It can take years, sometimes decades, to get new physics"

What's the Matter?

ATLAS was initially designed and built with specific purposes in mind. In the meantime, advances in the field have brought new questions to the forefront. As it enters adulthood and its working life, expectations for it have grown.

Seven teams of physicists have toiled for years over how to maximise the probability of identifying the particular events they are interested in. These physics working groups are interested in even more stringent tests of the Standard Model; identifying evidence for the existence of the Higgs boson and supersymmetry (SUSY); searching for 'exotics' like micro black holes predicted from versions of so-called 'string theory'; more comprehensive studies of the top quark; refining Monte Carlo methods; and the study of heavy ions which may give access to a new state of matter.

Once the LHC is running, our teams of sleuths will be sifting through the recorded data searching for physics events of specific interest. To draw on the oyster and pearl analogy one last time, these physicists will painstakingly open, with the specialised tools they have developed, their carefully selected oysters, hoping to encounter the particular pearls they so desire.

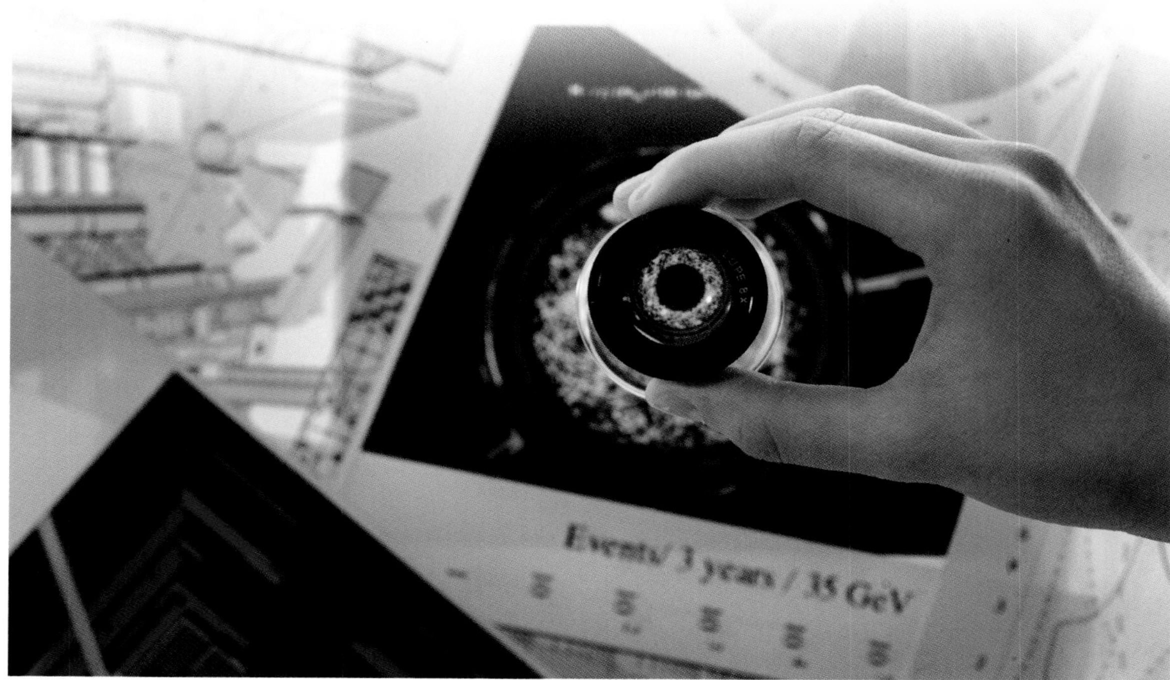

Matter of Fact

Earlier we took a plunge into the world of simulation and reconstruction. The last step in that process is analysis, when refined tuning of the selection criteria in each working group's software is put to the test with the real thing: event data. Once the flow starts, the Grid provides the power to run the complex analyses at remote sites around the world.

Physicists will begin by measuring the physical processes they already know as a way of understanding the detector and gaining confidence in its results. When something unexpected does emerge, they will look at the environment in which it was produced to see if they have discovered something new or simply do not fully comprehend the detector. New physics takes time to come about since nothing is published or disclosed until the physics community is nearly one hundred percent sure that what they have is, in fact, new and not a misreading. By the time an announcement is made, they will have scrutinised millions of raw events.

Higgs may fly

There are two complementary approaches to analysing data. The first one involves measuring everything and looking for inconsistencies. Should something not meet the expectations of known physics, it will be further investigated. The second approach is about seeking something specific like the Higgs. Theorists specify a particular form of new physics and then the experimentalists search for it.

99.9995 percent of the data that comes out of the collisions is thrown away as 'uninteresting background events' by the Trigger and Data Acquisition. For years to come, physicists will study the remaining data, elucidating some of the most intriguing questions we have about the building blocks of matter and generating a host more for the next generation of accelerators and detectors to answer.

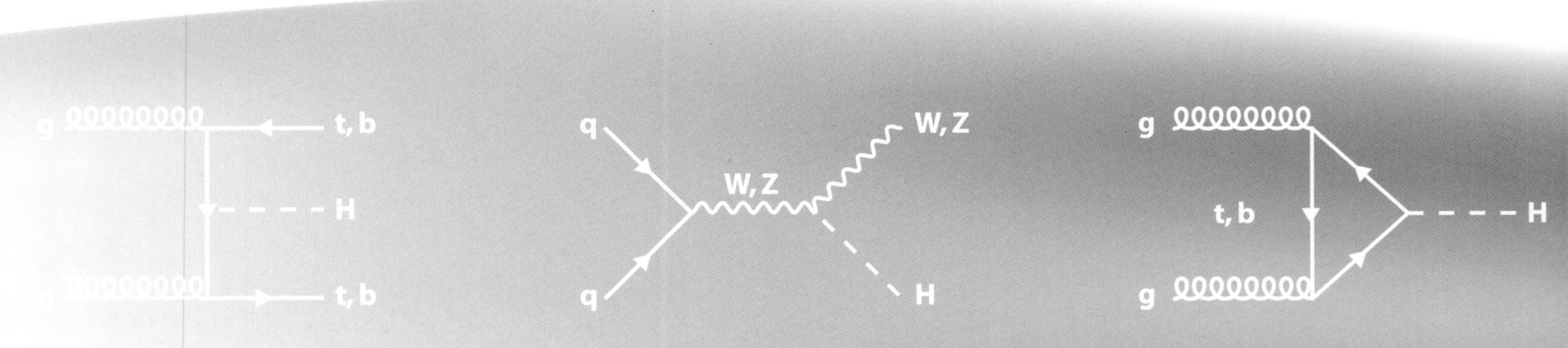

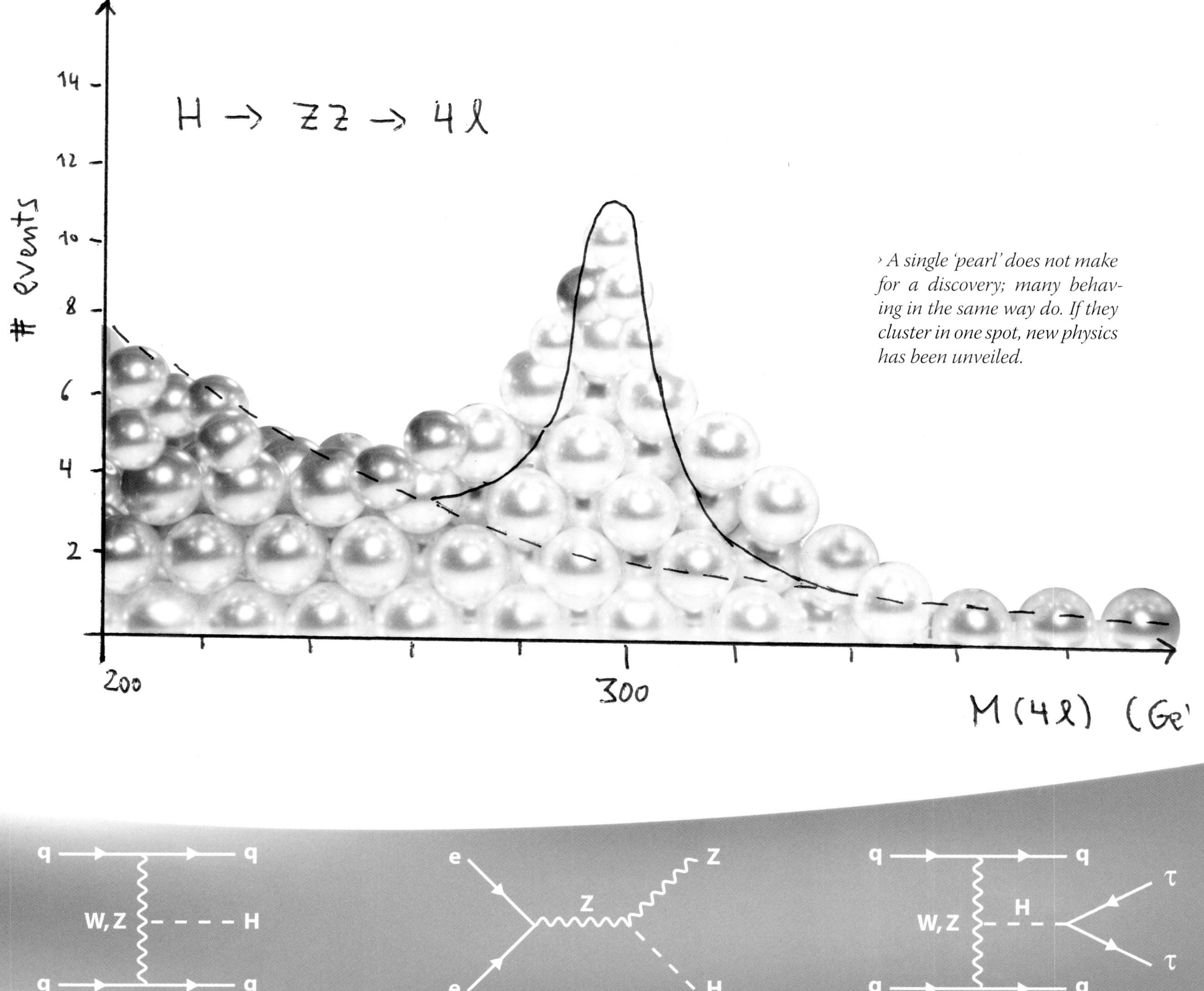

› *A single 'pearl' does not make for a discovery; many behaving in the same way do. If they cluster in one spot, new physics has been unveiled.*

› *Physics working groups brainstorm using simulation analyses, but soon their virtual world will be replaced by real data and long-awaited answers and confirmations.*

JUNGLE
D

Control Room to ATLAS

The light is muted, screens bright, hushed voices intermittent, the mood deliberate. Enter the control room and it is like stepping into a spy movie, the action revolving around the operations centre with its projected screens on the wall accessible to those in the know.

Fifteen work stations, each specialising in a particular aspect of ATLAS, working coherently, sit facing a wall where large screen shots of event displays and complicated charts are projected. Behind the work stations scientists intent on their numerous computer screens monitor the performance of the detector below ground.

The eye of the storm

Once ATLAS is running, the quality of the data can be checked in real time by taking an event sample and analysing it to ensure that the detector is performing correctly. The electronics are constantly being calibrated, the positioning of the muon chambers too. A slight change in temperature could impact the alignment of the muon chambers, for example, so the control room checks this.

Mission control

Each desk is backed by a satellite control room situated in barracks close by. Should something go wrong, a specialist is called upon. There is a distinct lack of chrome in the barracks, some offices carpeted and comfortable, others stark. Here the liquid argon specialist is on call 24 hours a day – many others too. Four to five hundred physicists will be needed each year to run the control room; the specialists will be fewer in number. Shift work looms ever closer.

› *The various desks specialise in particular aspects of the detector: monitoring the status of the gases, the running of the calorimeter, etc.*

SLIMOS and GLIMOS

Security at ATLAS is no laughing matter; the SLIMOS and GLIMOS take it very seriously indeed. When the GLIMOS (Group Leader in Matters of Safety) is away one of the SLIMOS (Shift Leader in Matters of Safety) gets to play in the control room checking that everything is normal down below. The ten computer screens in the security room are manned 24/7, 365 days a year.

15

"How far can we get doing science?"

› *Physics is often described in terms of elegance. Here, a simulated Higgs event against a plain background takes on a beauty all of its own.*

Going Out with a Big Bang

Long voyages come into their own when the unknown becomes a regular companion and chance meetings make for compelling exchanges. The quest ATLAS is about to embark on should fill in decades-long blanks, feed streams of research papers and generate a host of books. And so, as we reach the end of this book, the adventure truly begins.

Most journeys commence with the rough outline of an itinerary; this one is no different. On our odyssey through the fourth dimension we will hopefully encounter the Higgs under one guise or another, the more complex the better. At some point dark matter may pave the way to supersymmetry (SUSY), and extra-dimensions may become the next port of call. Some theories will disembark on the way, and unforeseen discoveries will become companions of fortune.

To get off to a good start, let us ask those in the theoretical know about their hopes and expectations for ATLAS and the LHC on the route ahead. Our distinguished travel companions are Lisa Randall, John Ellis and Peter Higgs.

Crash Course in Physics

We must first go back to the late 1800s, two centuries after Copernicus, to appreciate the advances that have been made in the field of physics. The Universe was perceived as being ordered sensibly, governed by Newtonian forces, the atom - Greek for 'indivisible' - the foundation of matter. Now, look back from where we stand today, and see just how far we have come.

Complacency was short-lived at the turn of the 20th century; X-rays, radioactivity and the divisibility of the atom had just been discovered, throwing the world of physics into turmoil. The 1930s saw the birth of the first circular particle accelerator. Today we have the LHC. Our knowledge of the building blocks of matter is still sketchy though our thirst for answers as healthy and insatiable as ever.

Warped mind

Lisa Randall has a plethora of questions she is seeking to answer, her ideas and research challenging the three dimensions of space we have lived in so far. She describes her work as model building: "I think of it as adventure travel through the world of ideas." She has been working on theories of extra dimensions for over a decade, finding that they could have astounding implications. She entered this world obliquely through the study of supersymmetry, and when she and her collaborator, Raman Sundrum, accidentally "stumbled" on the fact that space is warped, new dimensions opened up to them.

'Warped' multi-dimensional space-time could naturally explain a major puzzle in conventional particle physics: the extreme weakness of gravity relative to other forces. "Relationships between particle properties and forces that seem inexplicable when space is shackled to three dimensions appear to fit together elegantly in a world with more dimensions of space."

Her hope is that the LHC might just reach a high enough energy to spawn subatomic particles that have momentum in extra dimensions. "This momentum would be seen as extra mass," she explains. If any of these theories is right, ATLAS could be the bearer of such news. Evidence would include particles predicted from theoretical work by Kaluza and Klein, which leave traces of their existence in the dimensions we live in whilst travelling in extra ones. "These extra dimensions will not appear to us directly, but we'll know they exist because they have noticeable effects on matter inside them."

Anderson (PRD, 63')

$$\mathcal{L} = (D_\mu \phi)^+ D^\mu \phi - \frac{1}{4} F_{\mu\nu} F^{\mu\nu} - V(|\phi|)$$

$$D_\mu = \partial_\mu - ieA_\mu$$

$$F_{\mu\nu} = \partial_{[\mu} A_{\nu]}$$

Higgs (PRL, '64)

Englert & Brout (PRL, '64)

Higgs mechanism

$\oplus\ g \bar{\psi} \psi \phi$ (masses!)

$$\{Q^A_\alpha, \bar{Q}_{\dot{\beta}B}\} = 2\sigma^m_{\alpha\dot{\beta}} P_m \delta^A_B$$

(Wess & Zumino, NPB '73) ⟹ SUSY

Dark MATTER? ⟹ $\Omega_{CDM} \cong 0{,}25$

Dark ENERGY? $\Omega_b \cong 0{,}05$?

$\Omega_\Lambda \approx 0{,}7$

$\frac{\ell(\ell+1)C_\ell}{2\pi}$

Doppler peak

ℓ

SuperStrings???

"I hope ATLAS finds something genuinely new and we gain a better understanding of our world."

(Lisa Randall)

A known unknown

John Ellis's Eureka moment came walking down a corridor at CERN, when he thought to himself: "Ha! That's how you find the gluon." And it was. He explains his job as a theoretician as follows: "I think of things for the experimentalists to find and then hope they find something different." For over a quarter of a century he has been thinking of supersymmetry and the Higgs. "It would be nice to see them appear at some point," he smiles. However, he hopes to be astounded by the way in which they manifest themselves.

Ellis says of the Higgs that it is a known unknown. "I really believe the Higgs will be found at the LHC," he says, "but we don't know how it will manifest itself." He believes the Higgs will be "inadequate" though - "apologies to Peter" - and that it will require extra physics. "The Higgs has to have a small finite mass, but it's difficult to arrange without sticking in extra physics, and to my mind supersymmetry is the most natural example of extra physics."

Terra incognita

Supersymmetry is a beautiful mathematical theory, which at first did not seem to have much to do with physics. It was only in the 1980s that physicists realised it may help the Higgs do its job. "The Higgs boson gives particles their mass but has to have its own mass too, and by dropping supersymmetry in, its mass drops down to a sensible value." This theory, in which every known particle should have a heavier partner particle, could also help provide dark matter and is linked to string theory and the theory of everything (TOE) - whimsical acronyms are a way of life at CERN and one of Ellis's specialities.

"A lot of my research work is centred on trying to figure out how experimentally we try to detect these supersymmetric particles and in particular these dark matter particles." The existence of dark matter was first realised in the 1930s because of its gravitational effects on things we can see. It works like a 'glue' holding together the galaxies and clusters of galaxies. "In the early 1980s, some of us realised that the lightest of all these unseen supersymmetric particles has just the right properties to explain dark matter. I have high hopes of understanding it with the LHC - it is going places that no accelerator has gone before."

"As for ATLAS, if we don't find the Higgs because it's not there, it would be fantastic. It would overturn all the ideas of the past fifty years and be quite a breakthrough."

(John Ellis)

The nutcraker suite

We must cast our minds back to 1964 to find the germ of an idea that has grown into one of the main questions ATLAS is expected to answer. Half a lifetime later, and Peter Higgs will either come face to face with his namesake, or with something entirely different. "The work was done when I was 35 and I hope they get there before I'm 80."

"At the time I knew that what I'd discovered was important, and that it had potentially important implications, but I didn't know where it was to be applied. That only came a few years later." It came in the 1970s in fact, when an increasing number of physicists realised that this less 'trendy' strand of physics still held considerable potential and that there was work to be done, especially on weak interactions. The Higgs boson was first written about in 1976 in a paper on the Large Electron-Positron Collider (LEP), the LHC's predecessor. "Experimentalists were encouraged to look for a Higgs boson, which was difficult to do because it had no specific mass, and so the paper was rather apologetic because it was telling scientists to look for this boson on top of all the other work they had to do."

Then in the mid to late 1980s, what Peter Higgs refers to as the "relativistic version of the Anderson mechanism", became 'the thing' to look for. The Higgs had come into its own. "It's exciting to be so close today," he says, his eyes shining. And when asked what would happen if ATLAS did not find the boson, he smiles and says, "It needs to be there in one form or another, we wouldn't understand how to make sense of everything else without it." He has a bottle of bubbly ready to celebrate - let's hope he gets to pop it open very soon.

A space-time odyssey

Time to embark. Tickets please, space and time travel don't wait. Please ensure you have all your intellectual baggage with you, stow away all preconceptions and prepare to broaden your horizons.

› Peter Higgs shares his calculations of the Higgs mechanism that he calls the "relativistic version of the Anderson mechanism", also known as the Brout-Englert-Higgs mechanism, Higgs-Kibble mechanism or Anderson-Higgs mechanism. Earlier work by Yoichiro Nambu inspired the finding.

"Finding [the Higgs] is the end of an era. The more exciting thing is what comes after that, and ATLAS should open the way."

(Peter Higgs)

Credits

Photography Credits
Credit for the photographs in this book is given to Claudia Marcelloni,
© CERN and:

p. 102: © Alan Litke (Alan Litke and Alexander Sher)
p. 33 top: © Argonne Nat. Lab
p. 32 bottom center and right, p. 54: © Camille Moirenc
p. 79 top right and center left, p. 82 left: © Caroline Fabre
p. 32 bottom left, p. 35 bottom right: © CEA
p. 35 center top, p. 57 bottom right, p. 62 top, p. 85 top left, pp. 112, 118, 120, 122, 143: © CERN (Claudia Marcelloni and Max Brice)
pp. 26, 30 bottom right, p. 56 bottom left, p. 57 bottom center, p. 58 top and bottom center and left, p. 59 top,
p. 62 bottom left, p. 63 bottom right and center, p. 75 right, p. 78 top left and right and bottom right, p. 82 center and right, p. 84 bottom left, pp. 87, 88 bottom and top left, pp. 89, 117, 130, 131 bottom right and left, pp. 135, 141: © CERN (Max Brice)
pp. 30 bottom center, p. 33 bottom left and center, pp. 53, 59 bottom left and center, p. 109: © CERN (Laurent Guiraud)
p.137: © CERN (Mona Schweizer)
pp. 19, 55, 60 top, p. 111: © CERN (Patrice Loyez)
p. 81 bottom right: © Frank Taylor
p. 91 bottom center: © Guido Ciapetti
p. 56 bottom center, p. 57 bottom left, p. 78 center left and bottom left, p. 84 top and bottom right: © Heinz Pernegger
p. 60 bottom left: © IHEP, Protvino (Rinat Fakhrutdinov)
p. 91 bottom right: © INFN (Daniele Ruggeri)
p. 79 center right: © KEK, Japan (Takahiko Kondo)
p. 101: © Kenneth Cecire
p. 58 top right: © LAL_IN2P3/CNRS (Serge Prat)
p. 31 bottom left, right and center, p. 56 bottom and top right, pp. 77, 85 bottom right, p. 99: © LBLN (Roy Kaltschmidt)
p. 59 bottom right: © LIP, Lisbon (Agostinho Gomes)
p. 34 top: © Michael Schernau
p. 35 bottom center, p. 94 bottom left: © Michel Arnaud
p. 35 bottom left, p. 57 top, p. 60 bottom center and right, p. 75 left: © NIKHEF
p. 56 top left: © Norbert Wermes
p. 63 top and bottom right: © Peter Ginter
p. 158: © Robert Leslie
p. 58 bottom right, p. 88 bottom center and right: © Roy Langstaff
p. 48 bottom center: © STFC, UK
p. 33 bottom right: © Tumanov Y. A.
p. 31 bottom left: © Univ. Liverpool (Tim Jones)
p. 34 bottom left: © Weizmann Institute (Miki Koren)

Illustration Credits:
Credit for the illustrations in this book is given to:

P. 05, 42, 43, 46, 53 and 145: © Andre-Pierre Olivier
P. 75, 76, 106 and109: © Bernard Pirollet
P. 06: © CERN (Fabienne Marcastel)
P. 28 and 155: © CERN/LBLN (Joao Pequenao)

Glossary

THE STANDARD MODEL:
Every particle we know can be classified by a quantum property called 'spin' and is either a boson or a fermion. Bosons have integer values of spin, (0,1,2..) and fermions odd half values (1/2, 3/2,..). The currently favoured Standard Model of elementary particles is based on three 'flavours' or families of fermions of increasing mass. Each family contains one strongly interacting 'up-type' and one 'down-type' quark, together with one weakly interacting charged lepton (e.g. the electron) and one electrically neutral lepton (e.g. the electron neutrino). The forces between particles are due to interactions between fermions transmitted by 'force carrier' bosons such as gluons, photons, W and Z particles.

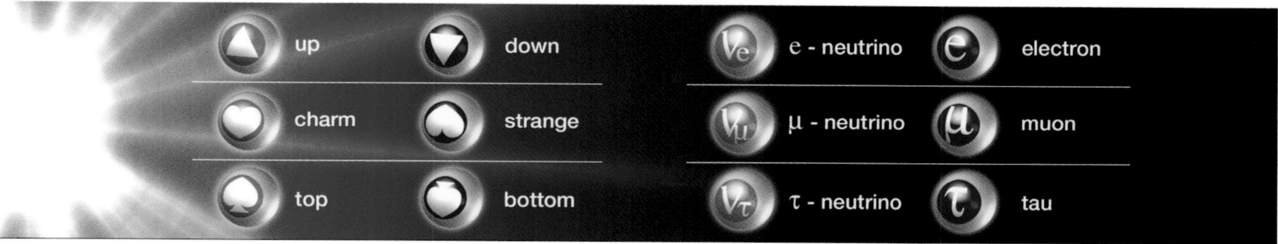

accelerator
A machine using electromagnetic fields to accelerate particles to speeds close to the speed of light. Typically, they are then magnetically steered into collision with either a fixed target or a counter-rotating, high energy beam.

antiparticle
Every fermion has an antiparticle with exactly the same mass but the opposite value of all other charges (quantum numbers). When particle meets antiparticle, they annihilate one another and their energy is available to create new particles.

ATLAS
A Toroidal Lhc ApparatuS, one of two general purpose detectors at CERN's Large Hadron Collider – the subject of this book.

atom
The smallest building block of elemental matter, consisting of a dense nucleus composed of protons and neutrons, surrounded by an electron cloud.

background
Unwanted type of event tending to obscure the 'signal' one is looking for.

beam
The particle stream produced by an accelerator, usually clustered in bunches.

Big Bang
A model of the creation and evolution of the Universe from an extremely dense and hot state 13.7 billion years ago.

black hole
A region where the gravitational field is so strong that not even light can escape.

boson
A 'force carrier' particle. The photon transmits the electromagnetic force, W and Z bosons the weak nuclear force and gluons the strongest, short-range nuclear force.

bunch
A collection of closely packed beam particles. Many bunches spaced at regular intervals form a particle beam.

calorimeter
A calorimeter measures the total energy deposited by particles in a substance. An electromagnetic calorimeter measures the energy deposited by electrons and photons; a hadron calorimeter does the same for hadrons.

CERN
Conseil Européen pour la Recherche Nucléaire, more commonly refered to as the European Laboratory for Particle Physics, located near Geneva, Switzerland.

charge
Particles with electric charge have electrical interactions. It is also possible to assign a 'strong charge' to each particle which has strong interactions.

collider
An accelerator in which two counter-rotating particle beams collide. The Large Hadron Collider can collide counter-rotating beams of protons or of heavy ions.

control room
The hub from which physicists operate a collider or detector system, such as the LHC or ATLAS.

cryogenics
The science and engineering associated with very low temperatures. Cryogens typically operate at temperatures below -180°C, and include materials such as liquid Helium, liquid Nitrogen and liquid Argon.

cryostat
A vessel used to contain very cold liquids or equipment submerged in those materials, such as liquid Helium or Argon.

dark matter
Matter that cannot be observed other than through gravitational effects.

decay
The transformation of a particle into two or more particles.

detector
A device used to sense and record the passage of a particle.

electromagnetic interaction
The interaction due to electric charge – one of the four fundamental types of interaction.

electron (e)
The most common lepton, with electric charge -1. The negative electric charge of an atom's electrons is balanced exactly by the positive charge of the protons in its nucleus.

element
All matter is composed of atomic species, elements, that are distinguished by the number of protons in their nuclei. So far, around 117 different elements have been observed.

eV/MeV/GeV
The energy gained by an electron which accelerates through a potential difference of one volt is one electron volt; a million eV is 1MeV, a billion is 1GeV.

event
The collision of two particles or a single particle decay. Also refers to the electronic record made by a detector of such a collision.

fermion
(Cf. boson above) A particle of odd-half-integer (1/2, 3/2, ...) spin. Quarks and leptons are the fundamental fermions, which exert forces on each other through interactions (strong, weak, electromagnetic and gravitational) mediated by boson force carriers (gluons, photons, W and Z bosons and gravitons) .

gluon (g)
The force carrier particle of strong interactions.

Grid computing
Massive data analysis on clusters of globally distributed computers.

hadron
A strongly-interacting sub-atomic particle composed of quarks, e.g. protons and neutrons, which are composed of three quarks each.

Higgs boson
The particle introduced theoretically to explain particle masses.

Inner Detector
The innermost part of ATLAS, inside the solenoid magnet, consisting of the Pixel Detector, the SemiConductor Tracker (SCT), and the Transition Radiation Tracker (TRT).

interaction
Response of a particle to a force due to another particle (as in a collision). In the Standard Model the fundamental interactions are the strong, electromagnetic, weak and gravitational interactions.

lepton
A fundamental fermion that is not composed of sub-particles and does not participate in strong interactions, e.g. the electron.

luminosity
A measure of the rate of collisions in a particle collider.

momentum
The product of mass and velocity. Conservation of the total momentum in an isolated physical system is a fundamental constraint in all particle interactions.

Monte Carlo
Computer simulation of interactions using large sets of random numbers to generate an approximate solution.

muon (μ)
The second heaviest charged lepton (in order of increasing mass).

muon chamber
Detector for registering muon tracks. Muons are among the very few particles to reach this outer layer of the detector.

neutrino (ν)
A lepton with no electric charge and therefore almost never directly detected in ATLAS. 'Missing energy' in a reconstructed event is indirect evidence for the presence of one or more neutrinos in the debris from a proton-proton collision.

neutron (n)
The electrically neutral component of an atomic nucleus. It is a hadron composed of three quarks, two down and one up.

noble gas
One of the chemically inert gases: Helium, Neon, Argon, Krypton, Xenon or Radon.

nucleus
The collection of neutrons and protons that forms the core of an atom (plural: nuclei).

photon (γ)
The force carrier particle of electromagnetic interactions.

pixel detector
Detector with 'pixellated' (miniature rectangular) electrodes.

postdoc
Post-doctoral researcher: a physicist after obtaining his/her PhD degree.

proton (p)
The nucleus of a hydrogen atom. It is a hadron composed of three quarks, two up and one down.

quantum
The smallest discrete, indivisible amount of any quantity (plural: quanta).

quantum theory
The theory describing physics on very small scales, where physical quantities are divided into discrete values (quanta).

quark (q)
A fundamental, strongly interacting fermion with electric charge of either 2/3 (up, charm, top) or -1/3 (down, strange, bottom) in units where the proton charge is 1.

resolution
Measure of the accuracy of a detector measurement, e.g. of energy or spatial position.

scintillator
A detector type in which a pulse of light is generated when a particle crosses it.

semiconductor
A solid with electrical conductivity between that of a conductor and an insulator. Silicon is the most common semiconductor, used in the ATLAS Pixel Detector and SCT.

solenoid
An electromagnet with a cylindrical coil giving an axial magnetic field.

spectrometer
A detector system that measures particle momenta using a magnetic field.

spin
A quantum unit of angular momentum. All elementary particles have either integer values or odd half-integer values of spin. The former are bosons, the latter fermions.

strong interaction
Responsible for binding quarks, antiquarks, and gluons to make hadrons such as protons and neutrons. The strongest, but shortest range of all the fundamental interactions.

superconductivity
Loss-free electrical conductivity achieved by some materials at cryogenic temperatures.

Supersymmetry
Theory in which every fermion has a (much heavier) boson partner, and every boson a (much heavier) fermion partner.

TeV
1 trillion electron Volts (1012 eV).

toroid
A type of magnet where the magnetic field forms the shape of a torus, or 'doughnut'.

tracking
The reconstruction of the path of a traversing particle in a detector.

Transition Radiation Tracker
A tracking detector specially adapted for highly relativistic electrons, consisting of many layers of narrow tubes ('straws'), each containing an axial wire and special gas filling. Traversing particles interact with the gas and surrounding media and create a signal in the wires.

Trigger/DAQ
The electronics and software to select, collect in real time and store the data from only the most interesting events for future analysis.

weak interaction
The interaction responsible for beta particle emission in radioactive decay by the interaction between quarks mediated by W bosons.

The ATLAS Collaboration

Acknowledgements

The ATLAS Collaboration is greatly indebted to all CERN departments and to the LHC project for their immense efforts not only in building the LHC, but also for their direct contributions to the construction and installation of the ATLAS detector and its infrastructure. We acknowledge equally warmly all our technical colleagues in the collaborating Institutions without whom the ATLAS detector could not have been built. Furthermore we are grateful to all the funding agencies which generously supported the construction and the commissioning of the ATLAS detector and also provided the computing infrastructure.

ATLAS acknowledges the support of ANPCyT, Argentina; Yerevan Physics Institute, Armenia; ARC and DEST, Australia; Bundesministerium für Wissenschaft und Forschung, Austria; National Academy of Sciences of Azerbaijan; State Committee on Science & Technologies of the Republic of Belarus; CNPq and FINEP, Brazil; NSERC, NRC, and CFI, Canada; CERN; CONICYT, Chile; NSFC, China; COLCIENCIAS, Colombia; Ministry of Education, Youth and Sports of the Czech Republic, Ministry of Industry and Trade of the Czech Republic, and Committee for Collaboration of the Czech Republic with CERN; Danish Natural Science Research Council; IN2P3-CNRS and Dapnia-CEA, France; Georgian Academy of Sciences; BMBF, DESY, and MPG, Germany; GSRT, Greece; ISF, MINERVA, GIF, DIP, and Benoziyo Center, Israel; INFN, Italy; MEXT, Japan; CNRST, Morocco; FOM and NWO, Netherlands; The Research Council of Norway; Ministry of Science and Higher Education, Poland; GRICES and FCT, Portugal; Ministry of Education and Research, Romania; Ministry of Education and Science of the Russian Federation, Russian Federal Agency of Science and Innovations, and Russian Federal Agency of Atomic Energy; JINR; Ministry of Science, Serbia; Department of International Science and Technology Cooperation, Ministry of Education of the Slovak Republic; Slovenian Research Agency, Ministry of Higher Education, Science and Technology, Slovenia; Ministerio de Educación y Ciencia, Spain; The Swedish Research Council, The Knut and Alice Wallenberg Foundation, Sweden; State Secretariat for Education and Science, Swiss National Science Foundation, and Cantons of Bern and Geneva, Switzerland; National Science Council, Taiwan; TUBITAK, Turkey; The Science and Technology Facilities Council, United Kingdom; DOE and NSF, United States of America.

The ATLAS detector design and construction have taken about fifteen years, and our thoughts are with all our colleagues who sadly could not see its final realisation.